TINKERING:
THE MICROEVOLUTION
OF DEVELOPMENT

Novartis Foundation Symposium 284

TINKERING: THE MICROEVOLUTION OF DEVELOPMENT

John Wiley & Sons, Ltd

This publication is designed to provide accurate and authoritative information in regard to
the subject matter covered. It is sold on the understanding that the Publisher is not engaged
in rendering professional services. If professional advice or other expert assistance is
required, the services of a competent professional should be sought.

Other Wiley Editorial Offices

John Wiley & Sons Inc., 111 River Street, Hoboken, NJ 07030, USA

Jossey-Bass, 989 Market Street, San Francisco, CA 94103-1741, USA

Wiley-VCH Verlag GmbH, Boschstr. 12, D-69469 Weinheim, Germany

John Wiley & Sons Australia Ltd, 33 Park Road, Milton, Queensland 4064, Australia

John Wiley & Sons (Asia) Pte Ltd, 2 Clementi Loop #02-01, Jin Xing Distripark, Singapore
129809

John Wiley & Sons Canada Ltd, 6045 Freemont Blvd, Mississauga, Ontario, Canada L5R 4J3

Wiley also publishes its books in a variety of electronic formats. Some content that appears
in print may not be available in electronic books.

Novartis Foundation Symposium 284

x + 289 pages, 40 figures, 2 plates, 4 tables

Anniversary Logo Design: Richard J Pacifico

British Library Cataloguing in Publication Data

A catalogue record for this book is available from the British Library

ISBN 978-0-470-03429-3

Typeset in 10½ on 12½ pt Garamond by SNP Best-set Typesetter Ltd., Hong Kong
Printed and bound in Great Britain by T. J. International Ltd, Padstow, Cornwall.
This book is printed on acid-free paper responsibly manufactured from sustainable forestry,
in which at least two trees are planted for each one used for paper production.

Contents

Participants

Rebecca Ackermann Department of Archaeology, University of Cape Town, Private Bag, Rondebosch 7701, South Africa

Jonathan Bard Department of Biomedical Sciences, The University of Edinburgh, Hugh Robson Building, George Square, Edinburgh EH8 9XD, UK

Michael A. Bell Department of Ecology and Evolution, Stony Brook University, Stony Brook, NY 11794-5245, USA

Paul M. Brakefield Institute of Biology, Leiden University, PO Box 9516, 2300 RA Leiden, The Netherlands

Graham Budd Uppsala University, Institutionen för geovetenskaper, Paleobiologi, Norbyv. 22, 752 36 Uppsala, Sweden

Robert Carroll Redpath Museum, McGill University, 859 Sherbrooke Street West, Montreal, Quebec H3A 2K6, Canada

James Cheverud Department of Anatomy and Neurobiology, Washington University School of Medicine, 660 S. Euclid Ave, St Louis, MO 63110, USA

Michael Coates Department of Organismal Biology and Anatomy, The University of Chicago, Culver Hall, 1027 East. 57th Street, Chicago, IL 60637, USA

Philip Donoghue University of Bristol, Department of Earth Sciences, Wills Memorial Building, Queen's Road, Clifton, Bristol BS8 1RJ, UK

Denis Duboule Départment de Zoologie et de Biologie Animale, Université de Genève—Sciences III, 30 Quai Ernest-Ansermet, CH-1211 Geneva 4, Switzerland

Brian K. Hall Department of Biology, Dalhousie University, 1355 Oxford Street, Halifax, Nova Scotia B3H 4J1, Canada

Benedikt Hallgrimsson Department of Cell Biology & Anatomy, Faculty of Medicine, University of Calgary, 3330 Hospital Drive NW, Calgary, Alberta T2N 4N1, Canada

James Hanken Harvard University, Museum of Comparative Zoology, 26 Oxford Street, Cambridge, MA 02138, USA

Jukka Jernvall Evolution & Development Unit, Institute of Biotechnology, University of Helsinki, 00014 Helsinki, Finland

Manfred Laubichler School of Life Sciences, PO Box 874501, Arizona State University, Tempe, AZ 85287-4501, USA

Daniel Lieberman *(Chair)* Departments of Anthropology and Organismic and Evolutionary Biology, Harvard University, 11 Divinity Avenue, Cambridge, MA 02138, USA

Gillian Morriss-Kay Weatherall Institute of Molecular Medicine, John Radcliffe Hospital, Headington, Oxford OX3 9DS, UK

Lennart Olsson Institut für Spezielle Zoologie und Evolutionsbiologie mit Phyletischem Museum, Friedrich-Schiller-Universität Jena, Erbertstr. 1, D-07743 Jena, Germany

Charles Oxnard School of Anatomy and Human Biology, The University of Western Australia, 35 Stirling Highway, Crawley, WA 6009, Australia

Elizabeth C. Raff Department of Biology and Indiana Molecular Biology Institute, G15 East Third Street, Bloomington, Indiana 47405, USA

Rudolf Raff Department of Biology and Indiana Molecular Biology Institute, G15 East Third Street, Bloomington, Indiana 47405, USA

Thomas Sanger *(Novartis Foundation Bursar)* Department of Biology, Washington University in St Louis, St Louis, MO 63130-4899, USA

David Stern Department of Ecology and Evolutionary Biology, Princeton University, Princeton, NJ 08544, USA

Irma Thesleff Institute of Biotechnology, PO Box 56, FIN-00014 University of Helsinki, Viikinkaari 9, Helsinki, Finland

Günter Wagner Department of Ecology and Evolutionary Biology, Yale University, 327 Osborn Memorial Laboratories, 165 Prospect Street, POB 208106, New Haven, CT 06520-8106, USA

Ken Weiss Penn State University, Department of Anthroplogy, 409 Carpenter Building, University Park, PA 16802, USA

Adam Wilkins Bioessays, 10/11 Tredgold Lane, Napier Street, Cambridge CB1 1HN, UK

The evolutionary developmental biology of tinkering: an introduction to the challenge

Daniel E. Lieberman and Brian K. Hall*

*Departments of Anthropology and Organismic and Evolutionary Biology, Harvard University, 11 Divinity Avenue, Cambridge, MA 02138, USA and *Department of Biology, Dalhousie University, 1355 Oxford Street, Nova Scotia B3H 4J1, Canada*

Abstract. Recent developments in evolutionary biology have conflicting implications for our understanding of the developmental bases of microevolutionary processes. On the one hand, Darwinian theory predicts that evolution occurs mostly gradually and incrementally through selection on small-scale, heritable changes in phenotype within populations. On the other hand, many discoveries in evolutionary developmental biology—quite a few based on comparisons of distantly related model organisms—suggest that relatively simple transformations of developmental pathways can lead to dramatic, rapid change in phenotype. Here I review the history of and bases for gradualist versus punctuationalist views from a developmental perspective, and propose a framework with which to reconcile them. Notably, while tinkering with developmental pathways can underlie large-scale transformations in body plan, the phenotypic effect of these changes is often modulated by the complexity of the genetic and epigenetic contexts in which they develop. Thus the phenotypic effects of mutations of potentially large effect can manifest themselves rapidly, but they are more likely to emerge more incrementally over evolutionary time via transitional forms as natural selection within populations acts on their expression. To test these hypotheses, and to better understand how developmental shifts underlie microevolutionary change, future research needs to be directed at understanding how complex developmental networks, both genetic and epigenetic, structure the phenotypic effects of particular mutations within populations of organisms.

2007 Tinkering: the microevolution of development. Wiley, Chichester (Novartis Foundation Symposium 284) p 1–19

'Tell you what though, for free, terriers make lovely fish. I mean I could do that for you straight away. Legs off, fins on, stick a little pipe through the back of its neck so it can breathe, bit of gold paint, make good . . .'

Pet Conversion Sketch, Monty Python's Flying Circus, Episode 10

In a favourite Monty Python sketch, an unscrupulous and imaginative pet salesman describes a few simple transformations by which he can convert a dog into a fish. The sketch is preposterous for many reasons, not the least of which

1

is the feigned biological naïveté of the customer. The pet store's unconventional practices also ridicule scenarios by which non-evolutionary processes might generate novel, adaptive forms. It has traditionally been believed that evolution occurs gradually within populations from the accrual of many incremental, tinkering-like transformations of existing structures. The radical transformation of a terrier into a fish is not only an evolutionary reversal but also absurdly un-Darwinian: 'As natural selection acts solely by accumulating slight, successive, favourable variations, it can produce no great or sudden modification; it can only act by short and slow steps (Darwin 1859, p 471).'

Darwin, of course, didn't know the mechanisms by which evolutionary change occurs, but he clearly believed in gradualism (Mayr 1991). The heart of his theory is that each generation within an interbreeding population inherits a range of phenotypic variations. Because of relentless competition caused by geometric rates of reproduction, limited resources, and/or environmental change, those variations that confer some benefit to an organism's chances of surviving and reproducing will increase in frequency. The accumulation of such variations over time leads not only to organisms that are better adapted to their environments, but is also assumed to be implicated in speciation—the most important type of evolutionary change.

Darwin and his contemporaries were working before modern genetics was born, and they struggled with problems of missing information and temporal scale. Interestingly, many of us continue to struggle with the same challenges even though we now have a much better grasp of the fundamentals of genetics. It requires little imagination to see how selection or other processes such as drift and founder effects can lead to different varieties of finches or tortoises; but it is considerably harder to understand how the same processes can lead from common ancestors to forms as diverse as whales and hippos, or even humans and chimpanzees. A part of this problem of comprehension lies in the nature of the fossil record, whose inadequacies lead inevitably to gaps. Most intermediate forms no longer exist, and (a bit) like Bishop Berkeley's conundrum of the falling tree, how can we understand a given transformation if we can't observe it? We are thus hampered by missing evidence and by a limited understanding of the mechanisms by which one form can transform into another. How does a jaw become an ear, a scale become a tooth, or a swim bladder become a lung? As we begin to understand more about the developmental processes that generate transformations leading to intermediate forms, we can begin to see a more complete and satisfying picture of what really happened in evolution and why.

The burgeoning field of evolutionary developmental biology (EDB) has had a major impact on our thinking about evolution above the level of the species (macroevolution). EDB has become a vital and exciting field by reopening the black box—so long ignored by evolutionary biologists—of the developmental bases for evolutionary change (see Hall 2003). At the same time, EDB has helped invigorate research on evolution within developmental biology. A key element in the origins of EDB was a set of discoveries in developmental genetics in a few model systems and their extension to non-model system organisms. About 20 years ago we remember the excitement of the first papers on homeobox gene regulation that started to fill the pages of *Nature* and *Science* (e.g. McGinnis et al 1984, Shepherd et al 1984, Dolle et al 1989, Kessel & Gruss 1990, Hunt et al 1991). Other key influences were Gould (1977), Raff & Kaufman (1983) and Atchley & Hall (1991), all of which were eye openers for palaeontologists and other biologists because they outlined more integrative approaches to addressing the evolutionary questions we wanted to study. Suddenly, it seemed possible for palaeontologists, long preoc-cupied with major trends in macroevolution above the level of the species, to test hypotheses about how evolutionary change occurred within populations (micro-evolution) and during speciation events. EDB has thus had widespread effects, even in fields remote from developmental biology. For example, in the field of human evolution, researchers have begun to rethink issues of homology, and to ask questions about the developmental bases for the transformations we observe in the fossil record (e.g. Lieberman 1999, Lovejoy et al 1999, Lieberman et al 2004, Pilbeam 2004, Hlusko 2004).

Not surprisingly, EDB itself is evolving rapidly. Many of EDB's most spectacu-lar and early advances focused on large-scale shifts in body plan. A typical research framework has been to compare the developmental mechanisms that underlie differences in development between two or more standard model organisms (chickens, mice, zebrafish, fruit flies and nematodes) in their phylogenetic context. These comparisons have led to many basic insights about the genetic and devel-opmental bases for major variations in animal body form. However, more and more organisms are now being studied in greater and greater detail and in taxa with closer and closer relationships. One useful consequence of this combination of increased breadth and depth is to permit comparisons of large-scale evolution-ary shifts with smaller-scale differences between closely related species. Given that natural selection occurs most fundamentally on individuals within populations, it is especially appropriate that researchers are increasingly studying the developmen-tal bases for the generation of variation within populations (e.g. Stern 2000, Brakefield 2003, Shapiro et al 2004, Frankino et al 2005). As EDB increasingly turns its focus to microevolution and the developmental bases of within-species variation, we can look forward to a new synthesis of genomics, EDB and popula-tion genetics.

Ironically, one issue persistently raised by the growing focus on microevolution within EDB and other fields is equivalent to the problem of scale, noted above, that confronted Darwin and his contemporaries: are the evolutionary developmen-tal bases of phenotypic change at the microevolutionary and macroevolutionary scales comparable in terms of kind or just degree? Does microevolutionary change occur via the same developmental processes that characterize the differences between distantly related model organisms? How useful is it to compare mice and fruitflies (or dogs and fish) if we wish to understand how evolution happens in the sense of the within-population variations of phenotype upon which natural selection acts?

Modelling microevolutionary transformations

In order to test hypotheses about the EDB of microevolutionary versus macroevolutionary events, one needs a framework to compare changes within populations and between species in terms of the transformational processes by which genotype generates phenotype. A good place to start is with three related issues about the relationship between developmental change and evolutionary change: the scale of evolutionary changes within populations, the relationship between genotypic and phenotypic variation, and the hierarchical nature of developmental pathways.

Tinkering

Evolutionary change occurs because phenotypic variation within populations is generated through random alterations to existing pathways or structures. This point has been made many times, including by Darwin (1859), but perhaps was made most clearly by F. Jacob's (1977) useful and brilliant analogy between evolutionary change and tinkering ('bricolage' in French). Unlike engineers who design objects with particular goals in mind based on *a priori* plans and principles, tinkers create and modify objects opportunistically by using whatever happens to be available and convenient. Similarly, heritable novelties upon which selection can act are generated only through the effects of mutations in the genome that lead to alterations in proteins and/or regulatory mechanisms that affect the developmental processes that influence an organism's phenotype. In the case of biological organisms however, tinkering occurs with no goal in mind. With the exception of the purposelessness of biological change, Jacob's analogy of evolution by tinkering is a particularly pithy analogy of how evolution generates novelty at multiple levels of development, and it helps explain several key emergent properties of evolutionary change such as integration, constraint and functionality. Tinkered things tend to work because they make use of pre-existing or easily modifiable functional components. As Jacob (1977) explicitly noted, tinkering explains why novel forms are often capable of developing, reproducing, and avoiding the fate of being hopefull monsters. And because of tinkering, all evolutionary change is constrained by the historical contingency of what happens to be available at given times in given lineages (Gould 2002).

Analogies tend to be dangerous forms of reasoning because, on inspection, they often break down in terms of utility and applicability. Jacob's tinkering analogy, however, has stood the test of time because it is so apt. Interestingly, and as noted by Duboule & Wilkins (1998), Jacob's essay is also one of the first clear expositions of the logic of EDB, but was written at a time when most evolutionary theory came either from systematists and paleontologists or population geneticists (see Mayr 1982). It took at least a decade for developmental biology to catch up with Jacob, but many of EDB's basic theoretical insights recall his analogy, explicitly or implicitly. One obvious example is the reuse of basic toolkit genes during development. As argued by Carroll et al (2001) and others (e.g. Hall 1999, Wilkins 2002), much if not most evolutionary change does not derive from new genes, but from new ways to deploy old genes in new contexts to generate novel forms. Tinkered developmental pathways typically alter phenotype by changing the timing/rate or site of expression of basic processes (leading to heterochrony or heterotopy, respectively). While these new pathways are likely to be successful because they use elements of proven function, they also lead to high 'workloads' for many basic toolkit genes that are expressed in many different contexts, and thus require elaborate *cis*-regulatory control (Duboule & Wilkins 1998). Another, related form of tinkering is the duplication and re-use of entire phenotypic modules that can take on novel functional roles (see Klingenberg 2005).

Transformations of genotype-to-phenotype

A second issue to consider is the complex relationship between genotypic and phenotypic variation. Natural selection acts within populations on phenotypically different individuals who vary in fitness. Since most phenotypic variation is complexly structured by many genes and by many developmental interactions, it follows that any theory of evolutionary change must be able to account for the relationship between genetic variation and the variation of complex phenotypes within populations.

This problem was highlighted succinctly by Lewontin's (1974) classic model of evolutionary change, shown in Fig. 1, which lays out the four sets of transformational 'rules' that generate evolutionary change in the relationship between genotype and phenotype. These four transformational processes are: (1) developmental transformations by which genotype becomes phenotype; (2) population-level transformations such as natural selection, founder effects and so on that lead to changes in gene frequencies within an interbreeding population; (3) transformations of the genotype during gamete formation such as mutation, segregation and recombination; and (4) transformations of the genotype caused by reproduction, such as fertilization biases, assortative mating and so on.

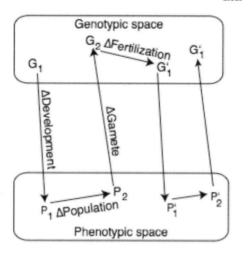

FIG. 1.   Lewontin's model of evolutionary transformations. See text for details.

Although Lewontin's model clarifies the relationship between developmental biology and population genetics, most population geneticists have focused on the last three of his transformations, and most developmental biologists have focused on his first transformation. A major and important exception to this division of labour was Atchley & Hall's (1991) model for the development and evolution of complex morphological structures, summarized in Fig. 2, which explicitly integrated developmental biology and population genetics. In particular, Atchley & Hall (1991) examined Lewontin's (1974) first transformation in the context of a particular empirical model, the mouse mandible. Atchley & Hall reasoned that complex structures such as the mandible initially comprise a finite number of semi-independent units (modules), each of which can be described by five parameters (the number of stem cells in the precondensation, the time of condensation initiation, the rate of cell division, the percentage of mitotically active cells, and the rate of cell death). Once initiated, these units then interact with each other through various epigenetic processes (e.g. induction) and through the pleiotropic effects of particular genes. In some cases, these units also interact with the environment (e.g. responses to mechanical loading or nutrition).

A number of key features made the Atchley & Hall (1991) model important in the origins of EDB. First, the model outlined how just a limited number of transformational processes by which genotype becomes phenotype can be used as parameters to study the evolution of complex structures. Second, by drawing on the work of Lande (1979), Cheverud (1984) and others, they explicitly formalized

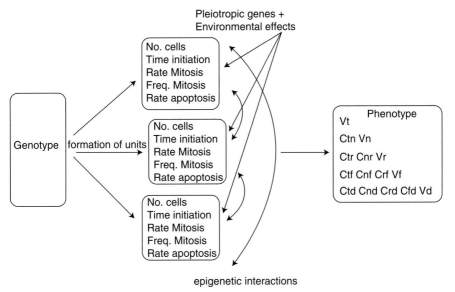

FIG. 2. A simplified view of the Atchley & Hall (1991) model whereby genotype generates units (here modelled in terms of skeletal condensations) whose size, shape and integration are influenced by a wide range of interactions. The end result of these interactions and processes is a particular phenotype characterized by a unique variance/covariance matrix. Changes to any of these generative steps will lead to predictable changes in phenotype as expressed not only in terms of heterochrony but also integration.

the effects of these changes in terms of patterns of variance and covariance. Finally, they also clarified how changes to developmental processes can be studied using heterochrony and/or allometry. Not surprisingly, the most profound impact of Atchley & Hall (1991) was on vertebrate morphologists, helping to spawn a rich literature on morphological integration (see Cheverud 2007, this volume). Unfortunately, Atchley and Hall's model was not applied as widely by geneticists and developmental biologists who work on plants, invertebrates and/or non-skeletal tissues. Part of this problem was that Atchley & Hall (1991) focused primarily on cellular processes relevant to skeletogenesis. It is much easier to study morphological integration in bones than in other tissues. In addition, Atchley and Hall mostly considered those cellular processes that regulate the size and origins of particular units (skeletal condensations), but did not discuss explicitly the genetic and developmental regulation of these processes. Thus, it has been a challenge for many developmental biologists to extrapolate Atchley and Hall's model to the particular tissues and/or developmental processes they study.

Regulatory hierarchies in development

A final issue to consider is how recent advances in our understanding of developmental pathways and their regulatory hierarchies help us clarify where and how selection can tinker with development to generate evolutionary change. This is a dauntingly large topic that has been the subject of many recent and diverse reviews (e.g. Gerhardt & Kirschner 1998, Hall 1999, Carroll et al 2001, Davidson 2001, Wilkins 2002, West-Eberhard 2003, Carroll 2005), but stepping back from the many details, one can make a few useful generalizations, illustrated in Fig. 3. Along the many steps by which genotype transforms into phenotype, there is a constant interaction between two interrelated sets of pathways: genetic and epigenetic (the latter defined, *sensu* Waddington, as interactions between a given gene and its environment, including the actions of other genes). Both genetic and epigenetic pathways are generally hierarchical, but in very different ways.

In terms of genetic pathways, there is a general hierarchical unfolding of connected gene activities, a network, that is characteristic of each developmental

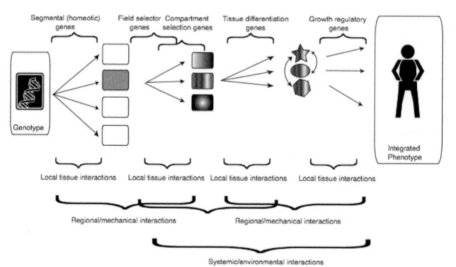

FIG. 3. A simplified model (based on *Drosophila*) of the interaction between genetic networks and epigenetic networks. These hierarchies occur concurrently and interactively. Note that the hierarchies depicted are arbitrary and not universally applicable; rather they are meant to illustrate the point that genetic networks combined with epigenetic interactions lead to integration at multiple levels of structure. See text for further details.

process. Carroll et al (2001) recently tried to summarize this sort of hierarchy for invertebrate development. While their scheme is overly simplified and not completely relevant to vertebrate development, it is nonetheless heuristically useful as a kind of pathway that may exist in certain cases. According to the Carroll et al scheme, *Drosophila* pathways typically begin with 'big gun' *patterning genes*, many of them homeobox genes, that define the basic units of the body such as the embryonic germ layers, neural crest, axial segments, and so on. Within these units, *field specific selector* genes then operate to trigger or constrain cascades of developmental events that help form major units such as organs or appendages. A now classic example is the eyeless gene, *ey*, which starts the development of eyes in *Drosophila*. Within these units, *compartment selector* genes then act to specify particular groups of cells that form axes (such as the way that the apical ectodermal ridge patterns the anteroposterior, dorsoventral and proximodistal axes of the limb). On top of these axes, *cell-type selector genes* then induce the differentiation of appropriate cell types to form the essential tissues of the body such as muscle, bone, or kidney. Finally, each of these cell groups can be subdivided into further units with differential rates and timing of growth, all of which are subject to regulation by specific *growth-related genes*. As noted above, other kinds of networks exist for different tissues and organisms (see Salazar-Ciudad & Jernvall 2004, for another example). The point is that genetic networks, by their very nature, tend to be hierarchical.

Genetic pathways—more appropriately considered networks because of their non-linear complexity—are not autonomous. Once started, they do not necessarily end up in the same place because a fundamental property of developmental pathways is that many, if not most, of the changes that occur during ontogeny are the result of interactions among different cells. These epigenetic interactions also have their own sort of hierarchy that runs interactively (and thus concurrently) with the genetic hierarchy described above (for a comprehensive review, see West-Eberhard 2003). At the local, most regionally specific level they include inductive interactions—sometimes reciprocal—between neighbouring cells that regulate the differentiation, proliferation and growth of particular cell lines, as well as influence their rate of migration and apoptosis. Classic and well-understood examples include the interactions between mesenchymal and epithelial cells in the face that regulate the formation of tooth germs (see Peters & Balling 1999, Jernvall & Thesleff 2000), or the interactions between mesodermal and ectodermal cells in the limb bud that initiate limb formation and regulate its patterning (see Tickle 2002). At a higher level, cells in a particular region are also influenced by signals such as diffusion gradients of morphogens (e.g. *Shh* in the limb bud). Cells in a given region can also be affected by mechanical and other environmental stimuli that, in turn, induce regional genetic and developmental responses. A good example is the way that brain growth up-regulates FGF2 expression in its surrounding membrane (the dura mater), which then up-regulates growth in sutures by turning on various transcription factors (e.g. MSX2 and TWIST) that stimulate osteogenesis (see Opperman 2000, Wilkie & Morriss-Kay 2001). Finally, cells within a given region or even the whole body are regulated generally by systemic endocrine factors, which, themselves, are regulated by a wide range of environmental and intrinsic stimuli.

Although the above model of developmental hierarchies summarized in Fig. 3 is necessarily rather general, and arguably too simplistic, it illustrates an important point about how tinkering can operate to generate potentially useful phenotypic change. Notably, genetic networks combined with epigenetic interactions at increasing levels of specification from the local to regional to organism levels, lead to integration at multiple levels of structure. Put differently, a key emergent property of increasing specification combined with regional and organism-level organization is a highly integrated phenotype. Change at any level, from mutations that affect the *cis*-regulation of how, when and where a particular gene is transcribed in a given cell line, to how much oestrogen is pumped throughout an organism's body, rarely lead to large-scale independent effects. Instead, their actions are modulated by numerous interactions at different levels. For example, mutations that lead to extra fingers or toes, typically lead to digits that have (more or less) an appropriate set of muscles, nerves and vessels to permit some degree of function.

Modelling the degree and rate of tinkering events

Let us now consider how the above insights can be used to model and test hypotheses about the degree and rate at which major transformations occur at the microevolutionary level. Recall that our basic question is whether the developmental differences we observe between distantly related organisms are comparable in type (their developmental and genetic bases) or merely in their degree of effect. In addition, do mechanisms of developmental change at the microevolutionary level provide us with any insights about whether speciations occur through gradual versus saltational transformations? In other words, can major transformations evolve rapidly with few or no intermediate states, and if so, are they similar to the differences we observe among distantly related organisms? EDB has rekindled interest in these questions by providing potential developmental support for the hypothesis that evolution can occur rapidly without intermediate transitional stages (e.g. Lovejoy et al 1999, Gould 2002, Raff et al 2003, Byrne & Voltzow 2004). The most extreme statement of this view, the hypothesis of punctuated equilibrium, posits that evolutionary patterns typically show long periods of stasis punctuated by rapid periods of saltational evolutionary change (Eldredge & Gould 1972, Gould & Eldredge 1993). Although the hypothesis was initially rooted in palaeontological observations, it potentially fits comfortably with EDB findings that distantly related organisms often use many of the same developmental genes and mechanisms, thereby potentially permitting a wide range of useful novelty to be generated (or lost) rapidly and without transitional forms via minor shifts in their regulation (Carroll 2005). As noted above, new segments can be generated by simple homeotic duplications; new appendages can be formed by heterotopic expression of existing field-specific selector genes (e.g. the much publicized *ey* mutants); and new tissues can be grown in new places by altered inductive interactions between neighbouring cell lines, many of which are changes in *cis*-regulation. Moreover, because developmental pathways necessarily take advantage of pre-existing mechanisms that generate integration, such shifts can lead to fully operational integrated organisms with new body plans rather than hopeless monsters. If major evolutionary transformations can and did occur via such simple shifts, then it follows that these shifts might have been rapid and saltational (that is, without many intermediate transitional forms). Although comparisons of distantly related organisms indicate that relatively simple developmental shifts can and do underlie major phenotypic differences, it does not necessarily follow that those changes occurred all at once without many intermediate transitional forms. As Darwin (1859, p 481) himself noted, one of the great challenges to thinking about evolution is to imagine the many transitional steps that can lead to large-scale changes: 'We are always slow in admitting great changes of which we do not see the steps'. Indeed, a number of arguments and observations support a more gradualist, transitionalist perspective whereby tinkering events typically generate small-scale changes with intermediate transitional forms. First, as noted above (e.g. Lewontin 1974), evolution occurs at its most basic level through the action of natural selection on individuals within populations. Thus, for organisms to remain part of interbreeding populations they cannot change so radically that their differences lead to reproductive barriers and/or isolation. In addition, since mutants must remain part of the gene pool in an evolving population (even one experiencing strong directional selection), any degree of reduced fitness in F1 backcrosses to the rest of the population and will select against change. Studies of hybridization generally support this point. Although distantly related species with distinct phenotypes (e.g. camels and llamas [Skidmore et al 2001]) can sometimes interbreed and produce fertile offspring, they rarely do. Even closely related and very similar species tend to have lower rates of reproductive success when they hybridize, leading to minimal introgression (Harrison 1993, Barton & Hewitt 1989). For example, geladas (*Theropithecus gelada*) and olive baboons (*Papio anubis*) regularly interbreed, but their F1 hybrids have low fitness (Jolly et al 1997).

   A second, more developmentally based reason to argue that evolutionary change is often if not usually gradual and transitional derives from the overlapping, interactive, and mutually-dependent genetic and epigenetic hierarchies illustrated in

Fig. 3. As noted above, the hierarchical nature of developmental pathways enables mutations of ultimately large effect to 'work', but the expression of such mutations is also modulated and constrained by processes of integration that are a fundamental property of developmental pathways. To use Waddington's (1957) terminology, the expression of a given mutation depends largely on its epigenetic landscape. The intrinsic effect of most mutations is neither big nor small, rather it is the genetic or environmental context of the mutation that determines its effect (see Wilkins 2007, this volume). One can therefore hypothesize, following Schmalhausen (1949) and Stern (1949), that mutations of large effect are more likely to be successful if their phenotypic consequences are initially constrained by their developmental milieu. Over time, via the effects of natural selection on other parts of the genetic network, the expression of such initially cryptic mutations is predicted to be manifested or even become enhanced as their regulatory control becomes modified through changes in integration (Futuyma 1987, Lauter & Doebley 2002). This kind of change presumably underlies the phenomenon of genetic assimilation whereby the expression of mutations occurs many generations after the mutation itself because of tinkering driven by environmental factors (see Palmer 2004).

Given these different scenarios, it is useful to compare alternative models for how evolutionary transformations at the subspecies or species level might generate evolutionary changes that vary in terms of their rate of transformation (saltational or gradual) and effect (small-scale or large-scale). Figure 4 attempts to summarize, in a highly simplified manner, several different kinds of pathways by which phenotypic units (squares) transform during ontogeny through various developmental processes (arrows). Because the number of units and their potential interactions with each other and the environment increase during ontogeny, the units in each pathway become increasingly integrated over time. However, the pathways differ in the degree of change that a given mutation can cause. At one of end of the continuum of possibilities are simple changes in development that have small-scale effects on a few aspects of the organism's phenotype (Fig. 4A). Such transformations, which can occur at different times during ontogeny (Fig. 4A illustrates one that is rather late), may be the most common type of evolutionary change. There are many examples, but one type that is especially well studied is the mutations to *Hox* expression in axial somites that generate variation in vertebral numbers and in the boundaries between vertebral types (e.g. Kmita & Duboule 2003). As shown by Pilbeam (2004), such mutations can account for the substantial variation in thoracic and lumbar vertebral numbers within hominoids, and may have played a role in the origins of bipedalism when modal number of lumbar vertebrae apparently changed from three in the last common ancestor of apes and humans to six in australopithecines. Other examples of this kind of small-scale tinkering include tandem repeat sequences in the *cis*-regulatory region of *Runx2* (an up-regulator of

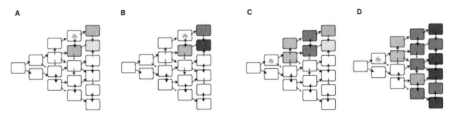

FIG. 4. Four models of evolutionary change. Boxes represent units of phenotype; arrows indicate developmental processes; fires represent mutations; ontogenetic time/stage progresses from left to right. In A, a mutation late in ontogeny leads to small-scale phenotypic change. In B, a mutation late in ontogeny leads to a larger phenotypic change, but with restricted effects on overall phenotype. In C, a mutation early in ontogeny leads to small-scale phenotypic change because of constraints imposed by other aspects of the genetic and epigenetic network of development. In D, a mutation early in ontogeny leads to wide-scale phenotypic change because of a lack of constraint by changes to the network.

osteoblasts) that apparently influence the length of the rostrum in dogs (Fondon & Garner 2004); and regulatory changes to *Pitx1* which leads to pelvic reduction in stickleback fish (Shapiro et al 2004).

A second possible type of transformation, illustrated in Fig. 4B, is a simple change in development that has large-scale effects on a few aspects of the organism's phenotype. Because such transformations have extensive effects with strong selective consequences, they might become rapidly fixed within a population by natural selection or other processes such as founder effects or drift, possibly causing rapid microevolutionary change. However such transformations, if they exist, are probably much more rare, and thus are hard to document. One possible example, which remains mostly untested, is the transformation from fins to limbs. Comparisons of distantly-related vertebrates such as mice and zebrafish suggested that the initial transition from fins to limb-like appendages most likely involved the recruitment of segments 9–13 in both *HoxA* and *HoxD* that, when expressed in the limb bud, generate an autopod with mobile wrists, ankles and digits (e.g. Shubin et al 1997, Coates et al 2002). The recent discovery of a fossil fish with a manus-like fin (*Tiktaalik roseae*) appears to confirm the existence of such a predicted transitional organism (Daeschler et al 2006, Shubin et al 2006), although it is not known to what extent the tetrapod manus is under extensive regulatory control, and how gradual the transition was in evolutionary time.

It is important to note, however, that neither of the hypothetical pathways described above in Figs 4A and B is highly integrated, and both model evolutionary changes in a few genes that have restricted effects. As is commonly appreciated, this sort of change is rare or unlikely because of extensive levels of pleiotropy, linkage and epistasis that constrain and modulate the expression of many mutations (Cheverud 1996). As a result, most organismal phenotypes exhibit substantial degrees of integration, typically manifested by high levels of correlation and covariation, as well as by various scaling relationships among their different components (see Chernoff & Magwene 1999, Dworkin 2005, Klingenberg 2005, Hallgrimsson et al 2005). Put differently, the widespread presence of extensive integration suggests that hierarchies of genetic and epigenetic pathways act as a sort of funnel to limit and structure the vast reservoir of genetic variation present in a population into a more restricted range of phenotypic variation. Evidence that developmental pathways restrict the generation of variation was recently demonstrated by an experiment by Hallgrimsson and colleagues, which compared the phenotypic effects of various mutations that influence the growth of different components of the skull in mice. Although the mutations themselves were completely different in terms of their primary effects (e.g. one acted on brain size, another on cranial base length), the different mutations led to similar patterns of integration, probably because their effects were structured by the many epigenetic interactions among the components of the skull that occur during development.

With these concerns in mind, a third more likely model, illustrated in Fig. 4C, is that most mutations of potentially large effect actually tend to have muted effects on phenotypic outcomes because of processes of canalization that are a fundamental property of most developmental pathways (Dworkin 2005, Klingenberg 2005). These processes, which include various stabilizing interactions and genetic redundancy, are adaptive (i.e. have been subject to selection) because they buffer organisms from the effects of major mutations. Consequently, they also lead to gradual rather than punctuated phenotypic change. Over time, however, selection may act on these pathways either to release the constraints they impose on development and/or to enhance the effect of the primary mutation (as shown in Fig. 4D). In such cases, we expect to see, eventually, widespread changes of large effect, but with intermediate, transitional forms.

Testing the latter two models is much more challenging because it requires knowing more about developmental pathways that is currently often the case. Knockout and knock-in experiments, however, provide useful evidence which support the constrained models illustrated in Figs 4C and 4D. As is well known, many are genes are identified and characterized because their knockouts have dramatic and widespread effects on phenotype. Typically, knockouts that affect coding regions of widely used transcription factors have especially pronounced and often lethal effects, while mutations that affect the transcriptional regulation of key genes can have large phenotypic effects in one genetic background but often produce much less of an effect when expressed in a different background (Pearson 2002). There are many examples of this phenomenon. To highlight one: knocking out the masterblind gene (*mbl*) in Zebrafish leads to an expanded jaw and reduced neural components in a TL background, but to much less expression in an AB background which partially rescues the phenotype (Sanders & Whitlock 2003). These and other similar cases (see Flatt 2005) indicate that mutations to complex pathways can be expected to lead to gradual or rapid evolutionary change depending on several factors, including the intensity of the selection pressure involved, and the degree of redundancy and constraint within the developmental pathway.

Indeed such effects are a predicted evolutionary outcome of tinkering itself. Over time, as tinkering events increasingly co-opt the same toolkit genes into more pathways, these genes take on new functions, and their combinatorial control becomes more complex (Kaufman 1987, Duboule & Wilkins 1998). Such complexity inevitably leads to more developmental stability and canalization, which means that many mutations that influence complex pathways may initially be unexposed to selection because they have minimal phenotypic effects. In other words, complex developmental pathways may be constrained to undergo more gradual change than simple systems, a phenomenon that Duboule & Wilkins (1998) term 'transitionism'.

Testing the models

Unfortunately, models of how tinkering events generate evolutionary change in populations produce more questions than answers. In order to resolve these questions, we especially need more information about how differences in developmental pathways generate variation within populations and between closely-related species (e.g. through QTL analyses on inbred lines and closely related species). For example, do mutations of large effect play a significant role in microevolutionary change, or are such changes simply the observed by-products of broad-scale comparisons? Moreover, to what extent and when are mutations of big effect (e.g. shifts in patterning) modulated and constrained by existing developmental pathways, both genetic and epigenetic? In addition, do phenotypic changes that lead to microevolutionary events typically occur from shifts at particular levels of development (von Baer's Law predicts they occur at later ontogenetic stages)? Finally, how do organisms in the same population cope with novel variations of integration and/or shifts in modularity (i.e., how developmentally different can a reproductively successful mutant be?).

Answering these and other questions will be an enjoyable challenge, one that requires a new synthesis of population genetics and developmental biology along the lines of Atchley & Hall (1991). Given the diversity and complexity of developmental pathways it is unclear if we will ever derive widely generalizable models applicable to a broad range of tissues and organisms. Our hunch, however, is that the developmental bases for most microevolutionary changes will follow Darwin's prediction of gradualism. As noted above, mutations with simple, direct effects on phenotype tend to "work" only when they result in minor phenotypic changes, whereas mutations of big effect are much more likely to lead to maladaptive organisms with low fitness that will be quickly removed from the gene pool. In reality, most developmental pathways tend to be complex and integrated, with many redundant steps, and to rely heavily on a small set of genes with heavy workload. Such complexity has, itself, evolved because it has enhanced fitness, and has permitted evolvability via tinkering. In such circumstances, natural selection favours mutations whose effects are buffered by complex, integrated pathways. One predicts these sorts of pathways to favour selection on aspects of pathways that release constraints and/or enhance the effects of chance mutations, thereby leading to gradual evolutionary change with intermediate phenotypes. When viewed over long time scales (as is typically the case when we compare distantly related organisms) we are seeing the cumulative effects of changes resulting in major shifts, but we are missing many of the complex changes in regulatory machinery that influence their expression.

Finally, although much of the work needed to better understand the developmental biology of microevolution will occur in the lab with model organisms such as mice and butterflies, it is useful to remember that such research has much broader implications. A particularly interesting challenge will be to test hypotheses about our own species' origins. In addition to satisfying our intrinsic interest in our own evolutionary history, several reasons make humans an exciting test case for a synthesis of EDB, population genetics and genomics. First, we have the complete human and chimpanzee genomes, along with extensive data on human genetic variation. In addition, we know from several lines of evidence that humans and chimpanzees shared a last common ancestor 5–8 million years ago that, phenotypically, must have been very much like a chimpanzee (Ruvolo 1997, Pilbeam 1996, Patterson et al 2006). In addition, we have a superb, well-studied and well-dated fossil record that extends back close to the estimated divergence time of apes and humans. And, finally, we have a rich knowledge of human developmental genetics from 'natural' knockout experiments in the form of various syndromes and diseases. Human and chimpanzee development will never be studied experimentally in the lab, but the above sources of data, combined with emerging new technologies, may help us figure out what genes changed in human evolution and how they were deployed. It may be funnier to imagine transforming a terrier into a fish, but it is far more interesting to decipher what processes actually transformed a chimpanzee-like last common ancestor into the earliest bipedal hominids, and thence through a series of transitions into modern humans.

Acknowledgements

We are grateful to Benedikt Hallgrimsson, David Oilbeam and Adam Wilkins for comments on a draft, and to the Novartis Foundation for making the symposium and this book possible.

References

Atchley W, Hall BK 1991 A model for development and evolution of complex morphological structures. Biol Rev Camb Philos Soc 66:101–157

Barton NH, Hewitt GM 1989 Adaptation, speciation and hybrid zones. Nature 341:497–503

Brakefield PM 2003 The power of evo-devo to explore evolutionary constraints: experiments with butterfly eyespots. Zoology (Jena) 106:283–290

Byrne M, Voltzow J 2004 Morphological evolution in sea urchin development: hybrids provide insights into the pace of evolution. Bioessays 26:343–347

Carroll SB, Grenier JK, Weatherbee SD 2001 From DNA to Diversity. Blackwell Science, Malden, MA

Carroll SB 2005 Endless Forms Most Beautiful: The New Science of Evo Devo. W.W. Norton. New York.

Chernoff B, Magwene PM 1999 Afterword. In: Olson EC, Miller RL (eds) Morphological integration. University of Chicago Press, Chicago, p 319–348

Cheverud JM 1984 Evolution by kin selection: a quantitative genetic model illustrated by maternal performance in mice. Evolution 38:766–777

Cheverud JM 1996 Developmental integration and the evolution of pleiotropy. Am Zool 36:44–50

Cheverud JM 2007 The relationship between development and evolution through heritable variation. In: Tinkering: the microevolution of development. Wiley, Chichester (Novartis Found Symp 284) p 55–70

Coates MI, Jeffery JE, Ruta M 2002 Fins to limbs: what the fossils say. Evol Dev 4:390–401

Daeschler EB, Shubin NH, Jenkins FA Jr 2006 A Devonian tetrapod-like fish and the evolution of the tetrapod body plan. Nature 440:757–763

Darwin C 1859 On the Origin of Species. John Murray, London

Davidson EH 2001 Genomic Regulatory Systems. Academic Press, San Diego

Dolle P, Izpisua-Belmonte J-C, Falkenstein H, Renucci A, Duboule D 1989 Coordinate expression of the murine Hox-5 complex homoeobox-containing genes during limb pattern formation. Nature 342:767–772

Duboule D, Wilkins AS 1998 The evolution of 'bricolage'. Trends Genet 14:54–59

Dworkin I 2005 Canalization, cryptic variation, and developmental buffering: a critical examination and analytical perspective. In: Hallgrimsson B, Hall BK (eds) Variation: A Central Concept in Biology. Elsevier, Amsterdam, p 131–158

Eldredge N, Gould SJ 1972 Punctuated Equilibria: An Alternative to Phyletic Gradualism. In: Schopf TJM (ed) Models in Paleobiology. SH Freeman, San Francisco, p 82–115

Flatt T 2005 The evolutionary genetics of canalization. Q Rev Biol 80:287–316

Fondon JW 3rd, Garner HR 2004 Molecular origins of rapid and continuous morphological evolution. Proc Natl Acad Sci USA 101:18058–18063

Frankino WA, Zwaan BJ, Stern DL, Brakefield PM 2005 Natural selection and developmental constraints in the evolution of allometries. Science 307:718–720

Futuyma D 1987 On the role of species in anagensis. Am Nat 130:465–473

Gerhardt J, Kirschner M 1998 Cells, Embryos & Evolution. Blackwell Science, Malden, MA

Gould SJ 1977 Ontogeny and Phylogeny. Harvard University Press, Cambridge, MA

Gould SJ 2002 The Structure of Evolutionary Theory. Harvard University Press, Cambridge, MA

Gould SJ, Eldredge N 1993 Punctuated equilibrium comes of age. Nature 366:223–227

Hall BK 1999 Evolutionary Developmental Biology, 2nd Ed. Kluwer, Dordrecht

Hallgrimsson B, Brown JJY, Hall BK 2005 The study of phenotypic variability: an emerging research agenda for understanding developmental-genetic architecture underlying phenotypic variation. In: Hallgrimsson B, Hall BK (eds) Variation: A Central Concept in Biology. Elsevier, Amsterdam, p 525–551

Harrison RG 1993 Hybrids and hybrid zones: Historical perspective. In: Harrison RG (ed) Hybid Zones and Evolutionary Process. Oxford University Press, p 3–12

Hlusko LJ 2004 Integrating the genotype and phenotype in hominid paleontology. Proc Natl Acad Sci USA 101:2653–2657

Hunt P, Gulisano, M, Cook M et al 1991 A distinct Hox code for the branchial region of the vertebrate head. Nature 353:861–864

Jacob F 1977 Evolution and tinkering. Science 196:1161–1166

Jernvall J, Thesleff I 2000 Reiterative signaling and patterning during mammalian tooth morphogenesis. Mech Dev 92:19–29

Jolly CJ, Woolley-Barker T, Disotell TR, Beyene S, Phillips-Conroy JE 1997 Intergeneric Hybrid Baboons. Int J Primatol 18:597–627

Kaufman SA 1987 Developmental logic and its evolution. Bioessays 6:82–87

Kessel M, Gruss P 1990 Murine Developmental Control Genes. Science 249:374–379

Klingenberg CP 2005 Developmental constrains, modules and evolvability. In: Hallgrimsson B, Hall BK (eds) Variation: A Central Concept in Biology. Elsevier, Amsterdam, p 219–247

Kmita M, Duboule D 2003 Organizing axes in time and space; 25 years of colinear tinkering. Science 301:331–333

Lande R 1979 Quantitative genetic analysis of multivariate evolution, applied to brain: body size allometry. Evolution 33:402–416

Lauter N, Doebley J 2002 Genetic variation for phenotypically invariant traits detected in teosinte: implications for the evolution of novel forms. Genetics 160:333–342

Lewontin RC 1974 The Genetic Basis of Evolutionary Change. Columbia University Press, New York

Lieberman DE 1999 Homology and hominid phylogeny: problems and potential solutions. Evol Anthropol 7:142–151

Lieberman DE, Krovitz GE, McBratney-Owen B 2004 Testing hypotheses about tinkering in the fossil record: the case of the human skull. J Exp Zool B Mol Dev Evol 302:284–301

Lovejoy CO, Cohn MJ, White TD 1999 Morphological analysis of the mammalian postcranium: a developmental perspective. Proc Natl Acad Sci USA 96:13247–13252

Mayr E 1982 The Growth of Biological Thought. Harvard University Press

Mayr E 1991 One Long Argument: Charles Darwin and the origin of modern evolutionary thought. Harvard University Press

McGinnis W, Levine MS, Hafen E, Kuroiwa A, Gehring WJ 1984 A conserved DNA sequence in homoeotic genes of the Drosophila Antennapedia and bithorax complexes. Nature 308:428–433

Opperman LA 2000 Cranial sutures as intramembranous bone growth sites. Dev Dyn 219:472–485

Palmer AR 2004 Symmetry breaking and the evolution of development. Science 306:828–833

Patterson N, Richter DJ, Gnerre S, Lander ES, Reich D 2006 Genetic evidence for complex speciation of humans and chimpanzees. Nature 441:1103–1108

Peters H, Balling R 1999 Teeth. Where and how to make them. Trends Genet 15:59–65

Pearson H 2002 Surviving a knockout blow. Nature 415:8–9

Pilbeam DR 1996 Genetic and morphological records of the Hominoidea and hominid origins: a synthesis. Mol Phylogenet Evol 5:55–68

Pilbeam DR 2004 The anthropoid postcranial axial skeleton: comments on development, variation, and evolution. J Exp Zool B Mol Dev Evol 302:241–267

Raff EC, Popodi EM, Kauffman JS et al 2003 Regulatory punctuated equilibrium and convergence in the evolution of developmental pathways in direct-developing sea urchins. Evol Dev 5:478–493

Raff RA, Kaufman TC 1983 Embryos, Genes, and Evolution: The Developmental-Genetic Basis of Evolutionary Change. Macmillan, New York

Ruvolo M 1997 Molecular phylogeny of the hominoids: inferences from multiple independent DNA sequence data sets. Mol Biol Evol 14:248–265

Salazar-Ciudad I, Jernvall J 2004 How different types of pattern formation mechanisms affect the evolution of form and development. Evol Dev 6:6–16

Sanders LH, Whitlock KE 2003 Phenotype of the Zebrafish masterblind (mbl) mutant is dependent on genetic background. Dev Dyn 227:291–300

Schmalhausen II 1949 Factors of Evolution. Blakiston, Philadelphia

Shapiro MD, Marks ME, Peichel CL et al 2004 Genetic and developmental basis of evolutionary pelvic reduction in threespine sticklebacks. Nature 428:717–723

Shepherd JCW, McGinnis W, Carrasco AE, De Robertis EM, Gehring WJ 1984 Fly and frog homoeo domains show homologies with yeast mating type regulatory proteins. Nature 310:70–71

Shubin N, Tabin C, Carroll S 1997 Fossils, genes and the evolution of animal limbs. Nature 388:639–648

Shubin NH, Daeschler EB, Jenkins FA Jr 2006 The pectoral fin of Tiktaalik roseae and the origin of the tetrapod limb. Nature 440:764–771

Skidmore JA, Billah M, Short RV, Allen WR 2001 Assisted reproductive techniques for hybridization of camelids. Reprod Fertil Dev 13:647–652

Stern C 1949 Gene and character. In: Jepsen GL, Mayr E, Simpson GG (eds) Genetics, Paleontology and Evolution. Princeton University Press, p 13–23

Stern DL 2000 Evolutionary biology: The problem of variation. Nature 408:529–531

Tickle C 2002 Molecular basis of vertebrate limb patterning. Am J Med Genet 112:250–255

Waddington CH 1957 The Strategy of the Genes. MacMillan, New York

West-Eberhard MJ 2003 Developmental Plasticity and Evolution. Oxford University Press

Wilkie AO, Morriss-Kay GM 2001 Genetics of craniofacial development and malformation. Nat Rev Genet 2:458–468

Wilkins A 2002 The Evolution of Developmental Pathways. Sinauer Press, Sunderland, MA

Wilkins A 2007 Genetic networks as transmitting and amplifying devices for natural genetic tinkering. In: Tinkering: the microevolution of development. Wiley, Chichester (Novartis Found Symp 284) p 71–89

Wilkins A, DuBoule D 1998 The evolution of 'bricolage'. Trends Genet 14:54–59

Tinkering: a conceptual and historical evaluation

Manfred D. Laubichler

School of Life Sciences, Arizona State University, PO Box 874501, Tempe, AZ 8527–4501, USA

Abstract. Francois Jacob's article 'Evolution and Tinkering' published in *Science* in 1977 is still the *locus classicus* for the concept of tinkering in biology. It first introduced the notion of tinkering to a wide audience of scientists. Jacob drew on a variety of different sources ranging from molecular biology to evolutionary biology and cultural anthropology. The notion of tinkering, or more accurately, the concept of *bricolage*, are conceptual abstractions that allow for the theoretical analysis of a wide range of phenomena that are united by a shared underlying process—tinkering, or the opportunistic rearrangement and recombination of existing elements. This paper looks at Jacob's analysis as itself an example of conceptual tinkering. It traces the history of some of its elements and sketches how it has become part of an inclusive discourse of theoretical biology and evolutionary developmental biology that emerged over the last 30 years. I will argue that the theoretical power of Jacob's analysis lies in the fact that he captured a widespread phenomenon. His conceptual analysis is thus an example of an interdisciplinary synthesis that is based on a shared process rather than a shared object.

2007 Tinkering: the microevolution of development. Wiley, Chichester (Novartis Foundation Symposium 284) p 20–34

Francois Jacob's paper 'Evolution and Tinkering' has become one of the classic papers of post-World War II (WWII) science (Jacob 1977). In it, one of the leading molecular biologists of his generation reflects on some of the most fundamental and general properties of living systems, and does so with characteristic French flair and erudition. The paper easily combines details from molecular biology with problems of evolution and development. But it also incorporates philosophical discussions about causation and scientific explanation as well as structuralist ideas borrowed from cultural anthropology. To a historian this paper thus offers an interesting window into the intersecting worlds of science, philosophy and anthropology in Paris during the 1970s (Jacob 1988, 1998, Dosse 1997, Morange 1998, Jacob et al 2002). To a theoretical biologist, however, Jacob's paper emphasizes the power of conceptual abstraction, innovation and synthesis.

In this paper Jacob argues that tinkering, introduced here by means of a comparison between an engineer and a tinkerer—one who 'always manages with odds and ends' (Jacob 1977)—is a better representation of nature's ways than any concept of planned design or teleology. But how does Jacob define 'better'? Reading through the lens of his cultural milieu, we can identify a post-colonial and post-1968 fascination with *bricolage* that was considered to be an answer to technocratic systems of power. However, while this might be true, it does not explain the appeal of the idea of 'tinkering' within biology.

Jacob's essay is itself a piece of conceptual *bricolage* that proposes a theoretical framework that to this day inspires biologists and other researchers and enables them to establish connections between sometimes widely different domains of science. To many, Jacob's article is an exemplar of an interdisciplinary synthesis. Its success lies in the fact that tinkering refers to shared properties between numerous natural processes within many different contexts. Tinkering, therefore, enables us to *integrate* these varied domains of nature based on these shared underlying principles. Jacob's conceptual analysis is sufficiently abstract—tinkering is introduced as a general phenomenon, rather than a specific mechanism—so that it is possible to build, or to tinker, a theoretical synthesis that begins with the general features of this process and then characterizes the specific details of individual instances. It therefore accomplishes on a conceptual level the same unification as a general mathematical model, such as the replicator equation or the Price equation does within the framework of dynamical systems (Price 1970, Page & Nowak 2002, Komarova 2004, Nowak & Sigmund 2004).

Tinkering in development—a brief history

Today, as evidenced for instance by this very symposium, tinkering is considered to be one of the central concepts in explanations of patterns of phenotypic evolution. We speak of 'a genetic toolkit of development' and 'regulatory evolution' (Davidson 2001, Carroll et al 2005, Davidson 2006) and the fact that differences in gene expression and new combinations of existing genetic and epigenetic elements rather than new genes are the causes of phenotypic variation and novelty is now common knowledge. Jacob clearly foresaw the role of tinkering in evolution and development for explanations of phenotypic evolution. He refers to the major 'transitions', such as the origin of multicellular organisms, as consequences of 'reorganization of what already existed' (Jacob 1977, p 1164). In that sense, evolutionary developmental biology and molecular biology are largely filling in the gaps of what for Jacob were still little known 'regulatory circuits that operate in the development of complex organisms' (Jacob 1977, p 1165).

Jacob's argument was inspired on the one hand by philosophical ideas about the nature of scientific progress, cultural anthropology and structuralist thought, and on the other hand by his first hand knowledge of the details of molecular regulatory mechanisms. Combining these ideas and applying them to questions of evolution and development then led him to the canonical insight about biological systems: 'it is always a matter of tinkering' (Jacob 1977, p 1165).

Jacob's Nobel-crowned authority and the fact that at that time *Science* still published lengthy philosophical and theoretical reflections helped to turn 'tinkering' into a scientific household concept. The initial response by the scientific community to the idea of tinkering was mixed; the concept made intuitive sense and could be used to illustrate general principles (the way Jacob himself did in his essay), but it took a while for it to become the focus of specific research programs and questions. Only in recent years do we find a significant number of papers that explicitly refer to tinkering. However, especially in the context of explanations of development and phenotypic evolution, 'tinkering' and related ideas have already had a long history before Jacob canonized the concept. In this paper I want to sketch a small number of episodes in this history and 'resurrect' a few scientists and ideas that deserve to be recognized in the scientific lineage of Evo Devo as well as in the history of 'tinkering.'

Jacob himself refers to Darwin and his recognition that life's reality is far more *bricolage* than perfection, although Darwin's argument that 'the wonder indeed is, on the theory of natural selection, that more cases of the want of absolute perfection have not been observed' (Darwin 1859, p 472, cited in Jacob 1977, p 1163) is also a brilliant rhetorical move to counter his critics. But in the context of Darwin's use of embryological and functional anatomical arguments—the conditions of existence and the correlation between parts, for instance—we can find a rather explicit recognition of tinkering or *bricolage* (Richards 1992). Tinkering and *bricolage* were also important in Ernst Haeckel's explanations of the biogenetic law. Ideally recapitulation would yield a linear sequence of ontogeny recapitulating phylogeny with further development and morphological differentiation largely a consequence of terminal addition. But because developmental time cannot be stretched at will and also because larval forms have to survive in whatever environment they live, deviations from this palingenetic or 'law-like' sequence were frequent. These so-called caenogenetic features of developmental sequences can be seen as instances of tinkering; organisms responded to those factors that challenged the underlying linear and law-like expression of development as recapitulation. They were able to do that based on the combinatorial possibilities and potentialities of the developmental system. But as deviations from the underlying norm, they are clearly an instance of *bricolage* (Gould 1977, Laubichler 2003).

While Haeckel is today mostly remembered as a popularizer and advocate of Darwin in Germany, and the main champion of monistic materialism, his friend and contemporary August Weismann has an altogether more solid reputation. The eminent historian of biology Fred Churchill has recently called him a 'developmental evolutionist' thus indicating Weismann's legitimate place in the history of Evo Devo (Churchill 1999). The main problem for Weismann was how it is possible to simultaneously provide a mechanistic account of determination and differentiation during ontogeny and for the constancy of the hereditary material. The latter problem was especially important in the context of widely held neo-Lamarckian beliefs. Weismann's solution included (1) the postulate of the separation of germ-line and soma and (2) a complex conceptual system based on a hierarchy of hypothetical functional particles that together make up the hereditary material. These included *Idants*, corresponding to observable cytological units; *Ids*, those units of evolution that combine to form the *Idants*; *Determinants*, parts of *Ids* that specify somatic traits and that were distributed among different cell types during development as units of embryological differentiation; and finally the *biophors* that actively interacted with cytoplasm to accomplish physiological differentiation. The whole system was purely conceptual, intended to capture what was at that time known about the behaviour of cells and chromosomes during development and reproduction. What is important for our purposes is that in Weismann's system the determining particles act in a combinatorial fashion. Again, different combinations and quantitative differences between existing units account for the observable phenotypic variations (Weismann 1875–76, 1892, 1904, Laubichler & Rheinberger 2006). Tinkering and *bricolage*, indeed.

While Weismann's system was largely conceptual—thus resembling the philosophical side of Jacob's argument—Wilhelm Roux was concerned with establishing a whole new experimental and 'mechanical' approach to explanations of development. The history of *Entwicklungmechanik* is relatively well documented, as far as the history of embryology has been told at all (Mocek 1974), here I only want to briefly discuss how the ideas of tinkering and *bricolage* have also been part of Roux' conception of development. Roux' interest was to account for development mechanistically, avoiding all references to specific organic forces. Physics and chemistry thus provided the reference frame for explanations of development. Roux noted that in the physical domain forces act homogeneously, that is they are always of the same kind and differ only in quantitative terms. And while homogenous forces can also be observed in an organic system, organic causes can also act in non-homogeneous ways; they display what Roux called '*komplexe Wirkungsweisen*' (complex effects) (Roux 1905, p 18). These can, for instance, be seen, in the genuine biological processes of inheritance, differentiation, assimilation, and regulation. Roux' solution of how to incorporate these complex effects into an overall framework of mechanistic causation was again based on a combinatorial perspective. As he put it, 'experience shows that a combination of several effects can lead to completely new kinds of effects' (p 17). Therefore the resulting conception of organic causation that explains observable qualitative differences between morphological forms is again one of tinkering and *bricolage*. A small number of basic causal factors combine in different ways to generate the diversity of organic forms and processes.

Roux' biophysical model of *Entwicklungsmechanik* did not explicitly include hereditary factors; in that way it was more limited than Weismann's conception, but it contributed enormously to the establishment of experimental biology. On the other hand, during the early decades of genetics—after Morgan's encounter with a white-eyed male—explanations of phenotypic variation and evolution were actually less sophisticated. The emphasis was on the transmission rules of characters and while these also displayed a combinatorial logic, especially when recombination was considered, causal explanations of development were initially not part of these models. However, Morgan-style transmission genetics was but one research program during the first half of the 20th century, albeit a domineering one. Other traditions, such as developmental physiological genetics, were both a continuation and a transformation of *Entwicklungsmechanik*. In this case the physiological orientation emphasized the dynamical and regulatory aspects of developmental processes. These included the functional roles of cellular, biochemical (molecular) and hereditary components of developing systems. Genes were not just seen as abstract factors of transmission, but rather as elements in a causal chain of events that generate phenotypic features. Uncovering genetic effects within these reaction chains was an ambitious goal of experimental zoology, which could initially only be considered because the preferred model organisms in these studies were two species of moth, *Lymantria* for Goldschmidt and *Ephestia* for Kühn (Geison & Laubichler 2001, Laubichler & Rheinberger 2004). But despite their advantages for experimental manipulation, research with these larger and expensive to keep organisms was extremely tedious. Later, when first *Neurospora* and later *Escherichia coli* emerged as the preferred experimental systems, research became more manageable. These new model organisms together with new experimental techniques largely shaped molecular biology and provided the immediate context and reference frame for Jacob's work and ideas.

But despite all these complications, it was Alfred Kühn who, together with Nobel Laureate Adolf Butenandt, first uncovered a complete reaction chain from a gene to a phenotypic trait. As this discovery is today often (wrongly) attributed to Beadle and Tatum, and since Kühn's conception of the physiological actions of genes in development has also many features of tinkering and *bricolage*, I will briefly sketch his experimental and theoretical contributions. Furthermore, Kühn was not only in a 'race' with Beadle and Tatum after they switched to *Neurospora* as an experimental system. During the 1930s Beadle had worked on *Drosophila* with Boris Ephrussi in Paris, and followed a similar approach as Kühn and his co-workers—a fact well known to all biologists in Paris after WWII (Kohler 1994).

Kühn initially worked on wing patterns of *Ephestia* but switched to eye colour mutants after discovering a recessive mutant with many pleiotropic effects. This opened a new research field as this and similar mutations were 'remarkable in their developmental physiological effects' (Kühn 1932, 1959 [1936]). The size of *Ephesita* allowed for transplantation experiments—a form of tinkering all by themselves— and this soon led to the observation that wild-type tissues, such as sexual organs, could restore the process of pigment formation in mutants. A hormone, called substance A, was initially suspected to be the critical substance as only tiny amounts of what was determined to be a diffusible substance were needed (Caspari 1933). Further research revealed that the reaction chain was even longer; a primary reaction within the cytoplasm was identified to take place before the circulation of the hormone A (Kühn et al 1935). Attempts to purify the substance followed and after many trials it was identified as kynurenine, a derivative of the amino acid tryptophan (Butenandt et al 1940). Tryptophan was no hormone. This finding led Kühn to reconceptualize the physiological reaction schema of gene action in development. No longer just a linear reaction chain with intermediates, Kühn now thought of two interacting chains: a 'gene action chain' and a 'substrate chain'.

In a seminal paper Kühn summarized the results of a decade of research as follows:

'We stand only at the beginning of a vast research domain. [Our] understanding of the expression of hereditary traits is changing from a more or less static and preformistic conception to a dynamic and epigenetic one. The formal correlation of individual genes mapped to specific loci on the chromosomes with certain characters has only a limited meaning. Every step in the realization of characters is, so to speak, a knot in a network of reaction chains from which many gene actions radiate. One trait appears to have a simple correlation to one gene only as long as the other genes of the same action chain and of other action chains, which are part of the same knot, remain the same. Only a methodically conducted genetic, developmental and physiological analysis of a great number of single mutations can gradually disclose the mechanism for realizing the hereditary dispositions (*das Wirkgetriebe der Erbanlagen*)' (Kühn 1941, p 258).

Tinkering, once again. The events of WWII interrupted this programme and by the time German science was functioning again, work in this area had shifted to molecular biology and *E. coli*. One leading center of research was Paris, which brings us back to our main hero. There is, however, one important addition to this story and it is of a more conceptual nature. Besides being a skilled experimentalist, Kühn was also an influential author. During the last decades of his long life he taught a regular course on developmental physiology at the University of Tübingen. These lectures were published as a textbook (Kühn 1955, 1965, 1971). The second edition has an explicit discussion of developmental evolution. Here Kühn argues that due to the harmonious and precarious balance between genes in development, any new phenotype has to represent a *'neues entwicklungsphysiologisches Gleichgewicht'* (a new developmental–physiological equilibrium). Most mutations, however, disturb this developmental balance; therefore they will be eliminated by natural selection. And, discussing evidence from transplantation experiments and embryological induction with amphibians, Kühn concludes that phenotypic changes are largely the result of regulatory changes and new combinations of developmental sub-processes (*Teilvorgänge*). Divergent phenotypes generally share a common repertoire of developmental mechanisms; only the regulation of these elements differs between them. He quotes Baltzer, that in instances of phenotypic evolution we find '*große Plastiziät und Kombinierbarkeit von Teilvorgängen*' (large plasticity and the ability to recombine sub-processes). Tinkering and *bricolage*, indeed. I do not know whether Jacob knew about Kühn's ideas, but there was at least one other Noble laureate who was inspired by Kühn's view of developmental evolution. As a student Christiane Nüsslein-Volhart read Kühn's lectures and decided not to become a biochemist. The rest is history.

Conclusion: tinkering and theoretical biology

Jacob's notion of tinkering has been one of the more successful conceptual innovations within the biological sciences. Like its contemporary—the 'selfish gene'—its appeal has not been restricted to the biological sciences alone (Dawkins 1976). As we do not have space to further explore these issues of reception and dissemination of ideas, I want to conclude with a brief investigation of the epistemological role of the concept of tinkering within theoretical biology. This might, at first, seem a rather odd question. Jacob argues as a molecular biologist reflecting on some general features of life. He does not present a strict formal or theoretical analysis; at most he gives us a heuristic and methodological prescription about how to think about some interesting and general features of biological systems. Furthermore, Jacob's scepticism towards formal or mathematical theories and models of biological phenomena is well known. So, in what way, then, can we connect tinkering to theoretical biology?

The answer to this question critically depends on how we view theoretical biology. In recent years a broader perspective on theoretical biology has emerged that emphasizes the importance of conceptual analysis and innovation for theory formation within the biological sciences, and does not equate theoretical with mathematical biology. Within such an expanded conception of theoretical biology four different aspects are commonly distinguished: (1) data management and analysis, i.e. the domain of bioinformatics; (2) mathematical models and simulations; (3) concept formation and conceptual analysis; and (4) theory integration (Laubichler et al 2005). The question then, is how does tinkering fit into this expanded conception of theoretical biology?

What we find is that tinkering is actually connected to all four domains of theoretical biology. Successful data analysis and management, and even more so data mining strategies, depend on clear guiding principles and theoretical assumptions.

Evo Devo provides one such perspective for rapidly accumulating data in developmental genetics. What began as a straightforward attempt to correlate molecular factors with phenotypic traits was soon complicated by the fact that no simple correlations exist. We have already discussed how our current understanding of the genetic basis of morphological evolution incorporates the idea of tinkering in the form of such concepts as 'regulatory evolution' and the 'genetic toolkit of development'. These concepts provide the basis for models that enable us to successfully incorporate molecular and developmental data into evolutionary biology. Based on these data, present day evolutionary developmental biologists—several represented here at this symposium—are developing conceptual and mathematical models of phenotypic evolution that include developmental principles. Either explicitly or implicitly all of them incorporate and emphasize a version of *bricolage* and tinkering (see, for instance, the papers in this volume and in related volumes by Schlosser & Wagner 2004, Callebaut & Rasskin-Gutman 2005).

Furthermore, Jacob's paper is itself a prime example of conceptual analysis. He arrived at the notion of tinkering by asking what single concept would best describe a wide range of empirical phenomena in molecular, developmental, and evolutionary biology. He then derived his specific interpretation of tinkering through an act of conceptual *bricolage*, borrowing on the one hand ideas from other scientific domains, while on the other hand establishing a clear conceptual dichotomy between tinkering and engineering-based design. Tinkering was thus introduced as an abstract concept, even though it was derived with an intention to capture concrete empirical results about molecular and developmental processes and their evolutionary significance. The challenge then has been to translate the concept of tinkering into empirical research programs. This did not happen immediately, but more recently, because of the accumulated data of developmental genetics and the theoretical framework of evolutionary developmental biology, we begin to see the emergence of concrete empirical research programs, such as the ones discussed in this volume. Jacob himself anticipated this delay when he noted that not many details about the regulatory gene networks were known in 1977.

Tinkering is also a concept that is essential for the theoretical integration of different empirical domains within biology. This integrative function is in many ways the ultimate goal of theoretical biology. One aspect of theory integration is the reduction of explanatory complexity—a reversal of the Borges dilemma where the map of the empire eventually became as large and complex as the land itself and therefore was rendered useless (Laubichler & Pyne 2006). In scientific explanations explanatory complexity can be reduced either in the form of ontological reduction—where a complex phenomena is seen as a consequence of the interaction of less complex parts—or by means of a shared underlying process that is isomorphic between different domains of nature. Tinkering falls under the later category. Jacob emphasized this dimension of tinkering when he suggested that it occurs across different biological systems. In the future, once individual research programs will have contributed more details about tinkering in biological, social, and cultural systems its integrative dimension will only become more prominent.

In conclusion, tinkering is indeed a core concept of current theoretical and evolutionary developmental biology that is increasingly also the subject of specific empirical research programs about the micro and macroevolution of development. The idea that development proceeds through opportunistic combinatorial rearrangements of existing modules has a long history within biology. Jacob's main contribution to this debate has been to coin a specific catching term—tinkering—for these phenomena that lend themselves to further elaboration within all four dimensions of theoretical biology.

References

Butenandt A, Weidel W, Becker E et al 1940 Kynurenin als augenpigmentbildung auslösendes agens bei insekten. Naturwissenschaften 28:63–64

Callebaut W, Rasskin-Gutman D 2005 Modularity: understanding the development and evolution of natural complex systems. MIT Press, Cambridge, MA

Carroll SB, Grenier JK, Weatherbee SD et al 2005 From DNA to diversity: molecular genetics and the evolution of animal design. Blackwell, Malden, MA

Caspari E 1933 Über die Wirkung eines pleiotropen gens bei der mehlmotte ephestia kühniella zeller. Wilhelm Roux' Archiv für Entwicklungsmechanik der Organismen 130:353–381

Churchill F 1999 August Weismann: A developmental evolutionist. In: Churchill F, Risler H (eds) August Weismann: Ausgewählte Briefe und Dokumente. Freiburg, Universitätsbibliothek 2:749–798

Davidson EH 2001 Genomic regulatory systems: development and evolution. Academic Press, San Diego

Davidson EH 2006 The regulatory genome: gene regulatory networks in development and evolution. San Diego, Academic, Burlington, MA

Dawkins R 1976 The selfish gene. Oxford University Press, Oxford

Dosse F 1997 History of structuralism. University of Minnesota Press, Minneapolis, MN

Geison G, Laubichler MD 2001 Reflections on the role of organismal and cultural variation in the history of the biological sciences. Stud Hist Philos Biol Biomed Sci 32:1–29

Gould SJ 1977 Ontogeny and phylogeny. Belknap Press of Harvard University Press, Cambridge, MA

Jacob F 1977 Evolution and tinkering. Science 196:1161–1166

Jacob F 1988 The statue within: an autobiography. Basic Books, New York

Jacob F 1998 Of flies, mice and men. Harvard University Press, Cambridge, MA

Jacob F, Peyrieras N et al 2002 Travaux scientifiques de François Jacob. Paris, O. Jacob

Kohler R 1994 Lords of the fly: drosophila genetics and the experimental life. University of Chicago Press, Chicago

Komarova NL 2004 Replicator-mutator equation, universality property and population dynamics of learning. J Theor Biol 230:227–239

Kühn A 1932 Entwicklungsphysiologische wirkungen einiger gene von ephestia kühniella. Naturwissenschaften 20:974–977

Kühn A 1941 Über eine Gen-Wirkkette der Pigmentbildung bei Insekten. Nachr Akad Wiss Gött Math-Phys Kl p 231–261

Kühn A 1955 Vorlesungen über Entwicklungsphysiologie. Springer, Berlin Heidelberg New York
Kühn A (1959 [1936]). Autobiographisches. Nova Acta Leopold 21:274–280
Kühn A 1965 Vorlesungen über Entwicklungsphysiologie. Springer, Berlin Heidelberg New York
Kühn A 1971 Lectures in developmental physiology. Translated by Roger Milkman. Springer, New York
Kühn A, Caspari E et al 1935 Über hormonale Genwirkungen bei Ephestia kühniella Z. Nachr Akad Wiss Gött Math-Phys Kl p 1–29
Laubichler MD 2003 Carl Gegenbaur (1826–1903): integrating comparative anatomy and embryology. J Exp Zoolog B Mol Dev Evol 300:23–31
Laubichler MD, Hagen EH, Hammerstein P et al 2005 The strategy concept and John Maynard Smith's influence on theoretical biology. Biol Philos 20:1041–1050
Laubichler MD, Pyne L 2006 The borges challenge in biology. Bioessays 28:768–769
Laubichler MD, Rheinberger HJ 2004 Alfred Kuhn (1885–1968) and developmental evolution. J Exp Zoolog Part B Mol Dev Evol 302:103–110
Laubichler MD, Rheinberger HJ 2006 August Weismann and theoretical biology. Biol Theory 1:202–205
Mocek R 1974 Wilhelm Roux—Hans Driesch. Jena, VEB Gustav Fischer
Morange M 1998 A history of molecular biology. Harvard University Press, Cambridge, MA
Nowak MA, Sigmund K 2004 Evolutionary dynamics of biological games. Science 303: 793–799
Page KM, Nowak MA 2002 'Unifying evolutionary dynamics.' J Theor Biol 219:93–98
Price GR 1970 Selection and covariance. Nature 227:520–521
Richards RJ 1992 The meaning of evolution: the morphological construction and ideological reconstruction of Darwin's theory. University of Chicago Press, Chicago
Roux W 1905 Die Entwicklungsmechanik ein neuer Zweig der biologischen Wissenschaft. Wilhelm Engelmann, Leipzig
Schlosser G, Wagner GP 2004 Modularity in development and evolution. University of Chicago Press, Chicago
Weismann A 1875–76 Studien zur Descendenztheorie. 2 Bände. Leipzig, Wilhelm Engelmann
Weismann A 1892 Das Keimplasma. Eine Theorie der Vererbung. Jena Gustav Fischer
Weismann A 1904 Vorträge zur Descendenztheorie. 2. Auflage. 2 Bände. Jena, Gustav Fischer

DISCUSSION

R Raff: I enjoyed your historical presentation, but I'd like to say that the latter part of it was incomplete and misleading. You seemed to derive Evo Devo as a branch of developmental genetics. I would say that it is not. Developmental genetics has provided important concepts and tools, but Evo Devo arises equally from comparative biology through the work of people like Gavin de Beer and others who worked through the 20th century on what are essentially Evo Devo problems without the modern technology but very much with the questions in mind, including tinkering. You also leave out the important thoughts from palaeontology, particularly having to do with rates and patterns of evolution seen in the fossil record.

Laubichler: I should have made this clearer: it was a conscious decision on my part to focus on only one of the elements in the history of Evo Devo. This was just one lineage that I could present in 25 minutes. There is more of an appreciation about the history of Evo Devo in the traditions of comparative anatomy and palaeontology, because those practitioners tend to have a better historical memory about their own discipline. In developmental genetics, however, the tendency to ignore predecessors is more prominent. I chose to say that even within that particular tradition leading to Evo Devo there is a long history that evolves around the notion of tinkering.

R Raff: I think the developmental biologists, starting with Roux, concentrated on development. They were interested in the machinery of development and how it works, and didn't think much about evolution until the resurrection of the Hox genes much later in the career of developmental biology.

Bard: That's not correct. The whole of Waddington's thinking integrated evolution and development. The epigenetic landscape was based around evolution as much as development. He at least was keeping both balls in the air simultaneously.

R Raff: I don't deny that. I'm trying to trace Evo Devo as a branch of developmental genetics. This is misleading because it leaves out so much. It leaves out Waddington, for a start.

Lieberman: Isn't it interesting that Waddington's term 'epigenetics' has a completely different definition for developmental biologists? The evolutionary aspect of his idea has been ignored by many developmental biologists.

Weiss: I think we need to make a distinction between natural selection and evolution. Natural selection almost implies tinkering, and Darwin's view of it, and his main reason for using the term, was to oppose creationism. The difference between creationism and natural selection in evolution is essentially contingency. The word 'tinkering' got lost because of Gould and Lewontin's famous Spandrels of San Marco paper, and their use of the word 'contingency' to describe this phenomenon of evolution, which displaced the term tinkering in popular science parlance. Is there a distinction between tinkering in the sense we are using the term and the whole idea of natural selection as a contingent history?

Laubichler: I disagree. If you look into the history of developmental biology, turning into developmental genetics, particularly through the more physiological models of the 1920s into the 1950s, there always was an evolutionary (pre Evo Devo) orientation. It might be a distinction between a continental European and American tradition. This is a complicated story. Interestingly, all those developmental biologists from Roux onwards didn't care that much about natural selection. This opposition between natural selection and tinkering is clear. They were interested in evolutionary questions and morphological evolution, but they thought that the fundamental difference was the origin of variation, which came from development and not necessarily from selection. Selection was considered to be important but more of a fine-tuning force. The more fundamental mechanistic force was in development.

Weiss: You can't fine tune something unless it is there.

Laubichler: They were more interested in creating the new. How do you account for that?

Hall: That is very much what Jacob was doing in his article. His entire argument was couched in terms of natural selection. It is a way that natural selection can act by combinatorial tinkering, and this aspect hasn't been picked up in the way that it should have been.

Hanken: You used a broader definition of tinkering than I would have expected. Can you identify a plausible model for the evolution of morphological features that is not tinkering?

Laubichler: This is an interesting observation, and it raises the question of how useful the concept of tinkering might be for guiding an empirical research program. You might find differences between the emphasis on combinatorial rearrangements as opposed to more ontogenetic sequences where you have something and you continue to refine it without fundamental rearrangements.

Hanken: Some of the models that you offered as examples have also been offered as explanations of the origins of higher taxa. I didn't know that this too is regarded as tinkering!

Laubichler: From the developmental biology tradition I sketched, the question is from something you can study experimentally and processes you can infer, how can you then incorporate them within an interpretive theoretical framework to account for morphological variation? There was no distinction made that there should be some different forces for macro and microevolution. In this respect, it would be logically consistent that if you have to rearrange existing elements of a developmental system, and this gets you novelty as well as variation, then it would account for pretty much everything. There wouldn't be special forces or other mechanisms.

Wilkins: Jacob was specifically contrasting ideas of tinkering with ideas of optimization in engineering. He was saying that evolution does not act like an engineer. In the evolutionary literature, very few evolutionists have claimed that evolution does act as an engineer. However, it is quasi implicit in some of the population genetics literature about optimization processes. These are over short terms, and are relatively minor adjustments. This was the contrast Jacob was trying to make. Listening to your paper it struck me that perhaps Jacob's main contribution was the term 'tinkering' itself. Everyone had been talking about tinkering for a century, but not knowing it, like the character in Molière's play who had been talking prose all his life without knowing it. This term focused peoples' attention on the process.

Laubichler: I agree. The term captures something, except that Jacob wasn't very successful in the short term. Jacob was to an extent characterizing something that seemed blatantly obvious but was so general and abstract that it didn't help people to turn it into a productive research program. The term hung out there and his paper was cited as an interesting piece of conceptual thinking or rhetoric, but it didn't have an effect. Interestingly, over the last several years this has changed and there have been more references to tinkering in the empirical literature. The variety of empirical research programs has reached a level where there is more of a perceived need for some conceptual unification, and then the term tinkering comes in handy.

Wilkins: All of the examples you gave of individuals in the past who have used the concept without using the term have come from a developmental biology background. I don't think there were any population geneticists in your set of examples. Is this an interesting split? Have the population geneticists not come to grips with this concept?

Wagner: To a degree it is related to gradualism. I don't see any split there. Population geneticists can't say anything about where the variation is coming from.

Weiss: There was a long-standing debate between mutationism and selectionism, in which in mutationism the idea was that most new things had to come by new mutations, and selectionism proposed that it is possible to choose and mould simply from what already exists.

Wagner: That seems to be such a silly distinction.

Weiss: I'm not defending it; it was a long-standing discussion. When we talk about combinatorial issues, this goes back to very early models of polygenes. The question will be, what do people mean by the variation they are talking about? This would then raise the question in what sense does evolution take place in populations? Why is it the population, since it is individuals that develop?

R Raff: Evo Devo hasn't been without a similar term: the popular term for a similar concept has been 'cooption'.

Liebermann: Cooption or combinatorial rearrangement are expressing a part, not a whole, of what I think Jacob was arguing. Different folks from different fields have focused in on different aspects. It was an obvious idea; Jacob didn't come up with anything new but expressed it very broadly and clearly. He used an analogy that happened to be apt. Typically, most analogies are dreadful: they break down rapidly and are quite dangerous to use. Jacob's analogy is potentially interesting and useful, and this is rare.

Laubichler: One reason why there might be a different emphasis in explanations of morphological variation and evolution within the field of developmental biology, is because once it is studied experimentally it becomes clear that there are few functionally relevant parts in the system There is nothing in population genetic theory that says stop at 35000 genes if you want to compose a phenotype. But to a developmental biologist the limitations of the number of factors involved in the causal processes that make up a phenotype are obvious. This is why I think there is more of an emphasis on tinkering and related concepts within the tradition of developmental biology than population genetics. Cooption is an interesting hybrid concept. It is one that is inspired by some notions of the developmental system, but conceptualized more by the population geneticists.

R Raff: It has been used comfortably by the Evo Devo field, but with a developmental concept. This is the idea of modular evolution and reuse of components. It is not as broad a definition as tinkering, but it provided a term that has been fairly encompassing and useful.

Hall: In terms of population genetics, presumably we would agree that random mutation is tinkering. If you then add on selection, is that still tinkering?

Bard: The assay of tinkering is the effect of mutation, not the mutation itself. If a mutation is silent and leads to no change in phenotype in the next generation, then no one can tell whether there has been any tinkering.

Hall: The mutation process itself is random. But when you add selection to that, can you still talk of this as tinkering?

Bard: You only recognize it because it has been selected.

Lieberman: Jacob would argue that tinkering is merely the emergent property of natural selection. Natural selection happens because it is tinkering.

Stern: You just said that tinkering is evolution by natural selection. When I first read Jacob's paper I was underwhelmed, both because this struck me as the most obvious explanation and because we already had terms for everything described in the paper. But listening to this discussion, I think that tinkering really is evolution by natural selection. It can't simply be mutation because we do have mutations that cause dramatic changes in the phenotype that we would not call tinkering and which would not survive natural selection. This can't be what we are talking about; instead, we are discussing the transitions that were allowed to survive, either because they were neutral or selected. Then we are talking about evolution by natural selection.

Wagner: Tinkering is not only important for understanding how genomic elements are rearranged. Graham Budd's work on functional integration is an important problem that can only be conceptualised in this context. It is also a concept that may help us understand evolution at different levels of organization. In particular, in the context of functional evolution. Functional continuity is implied in the idea of evolution: how do we explain functionally quite radically different life forms?

Budd: It is unfortunate that the term Evo Devo has been used to apply to a group of scientists who feel aggrieved that they were left out of the neo-Darwinian synthesis! As palaeontologists, we only make inferences about developmental systems, and rather incompetently at that. As this discussion has shown, the idea of tinkering is a rather diffuse one: it has been used to talk about massive re-arrangements caused by Hox gene transformations, and also low-level changes. One thing we could think about is the use we could have for tinkering in the future, after this meeting clarifies the concept. One thing that is lacking is that the Evo Devo tradition in general has not been over concerned with population processes. As you say, the idea of functional continuity is critical for allowing the genome to pass from one generation to the next. If you go back to the beginning of the 1990s and David Jacobs' work on Hox gene transformations (Jacobs 1990), there hasn't been much of a worry from the Evo Devo field about getting things through populations, allowing mutation, drift and selection to act on your varia-tion and propagating it to the next generation. It seems that we have been living in two parallel universes for a long time, and it's about time we put these two worlds together. Unfortunately, hardly anyone has talked to people from the other camp.

Hall: I think the time has come to lose the term 'Evo Devo', for some of these reasons. What Manfred Laubichler was emphasizing was the integrative biology for the 21st century we are looking for. I think this is the sort of thing that Jacob was concerned about. It is not all in genes being passed on as packages. Other things are happening here, such as emergent properties. I was surprised to hear Manfred use the term Evo Devo, because I think we are at the stage where this term is becoming unhelpful, because it keeps us in our camps.

Duboule: The community itself is still trying to define what Evo Devo is. Con-cerning tinkering, as the only French-speaking person here, I am not fully con-vinced that 'tinkering' is the right translation of *bricolage*. In François Jacob's mind, I assume, *bricolage* meant something that is done just on the spot; there is no plan to do it. However, a *bricolage* can be planed and well organized before its making. A reason why I think this idea has received little attention among orthodox evo-lutionists is because *bricolage* is something that can hardly lead to something perfect. It is an *ad hoc* solution that you live with. It works, but it is not what you would have done, should you have had the opportunity to think about what to do. This concept is extremely dangerous for an orthodox view of Darwinism. This is some-thing we should discuss more.

Reference

Jacobs DK 1990 Selector genes and the Cambrian radiation of Bilateria. Proc Natl Acad Sci USA 87:4406–4410

Tinkering: new embryos from old—rapidly and cheaply

Rudolf A. Raff and Elizabeth C. Raff

Department of Biology and Indiana Molecular Biology Institute, Indiana University, Bloomington, IN 47405, USA and School of Biological Sciences, University of Sydney, NSW 2006, Australia

Abstract. Marine embryos and larvae reflect distinct life histories and body plans from their adults, and are relatively simple in morphology and genetic regulation. Evolution of development to produce highly modified larval forms can be rapid among closely related species. We have studied the mechanisms by which the non-feeding direct-developing larva of the sea urchin *Heliocidaris erythrogramma* evolved from an indirect-developing feeding larva, the pluteus. *H. erythrogramma* diverged within 4 million years from its sister species, *H. tuberculata*, which has a typical pluteus larva. Radical evolution of *H. erythrogramma* early development allows it to reach metamorphosis in three to four days versus the several weeks required for the pluteus. We have used embryology, cross species hybrids, and manipulation of gene expression in embryos to dissect developmental changes and the genic controls that underlie these changes. Evolution of a new larval form resulted largely from several heterochronies in which conserved regulatory pathways are shifted in timing, producing new temporal relationships to other developmental events. Other changes in gene regulation also have contributed to rapid evolution of larval features, including the origin of unexpected and novel tissue identities that transcend changes within homologous features.

2007 Tinkering: the microevolution of development. Wiley, Chichester (Novartis Foundation Symposium 284) p 35–54

Tinkering with larvae

There are tensions in our understanding of evolution. We see a grand macroevolutionary sweep extending over millions of years. These are the dramatic changes of one complex form into another. However, we are limited to the direct study of microevolution, the smaller changes within populations over short time scales. We must connect these vastly different scales to fully explain evolution. Another tension arises from the peculiar mix of deep conservation and profound change that is so often observed in macroevolution. This goes back to the very concept of a phylotypic stage, and suggests that constraints of developmental processes are somehow involved in evolutionary conservation of shared features of ontogenies. The phylotypic stage has been shown to vary, and sufficiently so as to have its reality questioned (Bininda-Emonds et al 2003), but changes to highly conserved defining basic body plan features do evolve far more slowly than major changes within body plan (e.g. for vertebrates, features such as novel tetrapod appendages, human brain, locomotor systems, body coverings, and endothermic metabolism). The experimental question is thus to find ways of asking how novel features arise within a relatively short time period in closely related taxa.

The time required for microevolutionary changes in *cis*-acting control regions of genes can be surprisingly short. Stickleback fish populations in lakes have diversified in morphology in the past 10 000 years. A small number of loci account for most phenotypic variation, with one gene, encoding the transcription factor Pitx1, controlling evolutionary pelvic reduction (Shapiro et al 2004).

The evolution of marine larval forms offers a spectrum of rapidly evolving larval features and provides an ideal system for study. The advantages of studying the evolution of novel features in marine larvae include morphologies and cellular complexities lower than in adults, simpler gene regulatory systems, phylogenetic information giving polarity of evolution, and rapid evolution of striking life history and morphological traits. Notable examples include annelid poecilogony (production of more than a single larval type within a single species [Gibson & Gibson 2004]), as well as dramatic developmental mode evolution among closely related species of mollusks, ascidians, starfish and sea urchins. In these, feeding planktotrophic larval forms have transformed into morphologically distinct viviparous, or brooded direct-developing, or planktonic non-feeding direct developing forms (Collin 2004, Jeffery et al 1999, Hart et al 1997, Byrne 2006, Zigler et al 2003). Losses of larval features involved in both planktonic feeding and appearance of novel features have occurred.

Studies of fossils that reveal mode of development in a sea urchin clade show repeated evolution of free living and brooded direct-developing larvae from feeding larvae. The non-feeding larvae had shorter species durations than planktotrophic species (Jeffery & Emlet 2003, Jeffrey et al 2003). Rates of evolution of novel larval forms are rapid, with estimates of evolution of viviparous starfish larvae from planktotrophic ancestors in as little as one to two million years. Evolution of direct developers has taken place in asterinid starfish within about four million years (Hart et al 1997). The evolution of direct-developing sea urchins from the feeding pluteus has taken place within four million years in *Heliocidaris*, and four to seven million years in temnopleurid sea urchins (Zigler' et al 2003, Jeffery & Emlet 2003).

A novel larval form

There has been an enormous body of functional developmental biology carried out with indirect-developing echinoids over several decades. This body of information makes it possible to study molecular mechanisms of developmental evolution in echinoids through comparisons of developmental regulatory processes to those of closely related indirect-developing sea urchin species. We are using this approach with two congeneric Australian sea urchins in the genus *Heliocidaris* (Raff 1996). The two species diverged about four million years ago (Zigler at al 2003) and in this short time *H. erythrogramma* has evolved a radically modified mode of development (Fig. 1A, B). *H. tuberculata* is a typical indirect-developer that retains the ancestral mode of development; this species has a 90 μm egg that develops into a typical feeding pluteus larva, which takes about six weeks to generate the juvenile adult sea urchin that emerges upon metamorphosis. *H. erythrogramma* has a 430 μm egg and produces a highly modified larva with a novel morphology, novel ectoderm, and a number of developmental heterochronies that have produced a blending or mosaic of embryonic and adult developmental processes.

Cross species hybrids

The harmonious development of cross species hybrids revealed an unexpected robustness of the joint operation of two genomes in a common cytoplasm (Raff et al 1999, Nielsen et al 2000). The ontogeny of the hybrids generated from fertilization of *H. erythrogramma* eggs by *H. tuberculata* sperm (*He* × *Ht*) is distinct from either parent (Fig. 1C, D), and indicates novel regulatory outcomes that are highly dependent on the cytoplasmic organization of the host egg. The fact that the *He* × *Ht* cross supports a well-integrated novel ontogeny that leads to metamorphosis into a sea urchin indicates that gene regulatory systems of the two species are similar enough to respond to each other and to upstream controls in an integrated way, notably those involved in the specification the dorsal-ventral (D-V) and left-right (L-R) axes. A key feature in *H. erythrogramma* is that these embryonic axes have become maternally established, rather than zygotically as in the indirect-developing species. The importance of the reorganization of maternal contributions is reflected in the converse cross of *H. tuberculata* egg fertilized by *H. erythrogramma* sperm (*Ht* × *He*): the reciprocal hybrids fail to form larval axes and die at late gastrula. Other cross species hybrid studies show that *H. erythrogramma* development has evolved rapidly whereas pluteus development has evolved slowly (Raff et al 2003). Other hybrids also show significant convergence in independently derived direct development in other sea urchins.

Mechanisms of rapid evolution

H. erythrogramma has been substantially modified in ontogeny relative to the planktotrophic larval form from which it is derived, as exemplified by *H. tuberculata*. Many aspects of the maternal contribution to development have been remodelled.

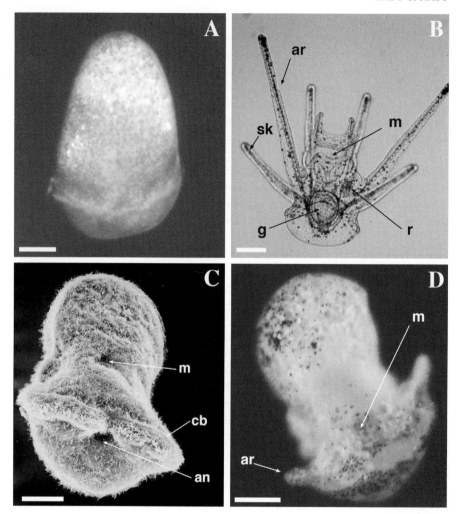

FIG. 1. Plasticity of larval form in the sea urchin genus *Heliocidaris*. (A) Three day-old *H. erythrogramma*, ventral view in bright field. The ciliary band is visible in the lower part of the larva. (B) 17 day-old *H. tuberculata* 8-armed pluteus ventral (oral) view in bright field. (C) Two day-old *He* × *Ht* hybrid larva in scanning electron microscopic view, ventral (oral) aspect. (D) Three day-old *He* × *Ht* hybrid, ventral (oral) view in bright field. Abbreviations: an, anus; ar, arm; cb, ciliary band; g, gut; m, mouth; r, rudiment of juvenile adult forming on left side of the larva; sk, skeletal rod, part of the calcareous larval skeleton of the pluteus. Scale bars = 100 μm. From Raff et al (1999) *Development* 126:1937–1945.

The first is in maternal provisioning of the 100-fold greater egg volume and changes in maternal organization of the egg (Villinski et al 2002). The primary embryo axes are established maternally (Henry & Raff 1990). In contrast, in indirect-developing sea urchins the animal-vegetal axis is determined maternally, but the oral-aboral (D-V) and L-R axes are only established within the embryo (Duboc et al 2004, 2005). Thus, major evolutionary changes leading to the direct pathway of development in *H. erythrogramma* reflect reorganizations of programs in oogenesis that are then played out in the novel form of embryogenesis.

The changes in the maternal program are a major factor in organization of the new programme of ontogeny. These differences can equate with differences in features of embryos generally thought characteristic of class or phylum taxonomic levels. However, evolutionary change to direct development is not sex limited; substantial changes have also occurred in spermatogenesis in *H. erythrogramma* relative to *H. tuberculata* (Table 1; Raff et al 1990, Smith et al 1990). Increase in sperm head size, partially but not solely due to increased DNA content, may reflect the consequences to fertilization processes associated with increased egg size, such as the requirement for the sperm acrosomal process to penetrate the increased thickness of the egg jelly coat (Raff et al 1999, Zigler et al 2003). There is a tight correlation between the switch to direct development in echinoderms with increased egg volume, increased sperm size, and increased genome size (Raff et al 1990, Wray 1996).

In the *H. erythrogramma* embryo, the cell fate map is highly modified from those of indirect-developing sea urchins, and cell cleavage patterns are modified such that there are no micromeres (Wray & Raff 1990). That this change has taken place within the divergence of two species shows that the high degree of evolutionary stability of the pluteus larval form does not preclude rapid changes under selection.

The large effects observed in *H. erythrogramma* pose a paradox. The molecular mechanisms that execute axial differentiation are conserved, but the programmes in which they are utilized have become reorganized such that the outcomes have changed. In both indirect developers and *H. erythrogramma*, the animal-vegetal (A-V) axis is determined in the egg. Expression of the A-V axis in *H. erythrogramma* early development is conserved and uses the same Wnt8 signal transduction pathway as described for indirect development (Angerer & Angerer 2003, Kauffman et al 2003).

In indirect developers, the D-V and L-R larval axes are initially set up through action of the Nodal pathway and involve, among other transcription factors, goosecoid (gsc) (Angerer et al 2001, Duboc et al 2004, 2005, Flowers et al 2004). Nodal is expressed first on the ventral side of the blastula and creates a signaling gradient that establishes the dorsal-ventral axis (Duboc et al 2004). Subsequently,

TABLE 1   Events in *H. erythrogramma* larval evolution and inferred mechanisms of change

Developmental Feature	Change	Apparent Mechanism
Gametes		
Egg		
Egg size	large increase in volume	heterochrony
Egg lipids	increased amount; change in composition	rate/duration loading, metabolic regulation
Axis determination	embryonic to maternal determination	heterochrony
Sperm		
Sperm size	increase in sperm head volume	increase in total DNA content (genome size); other unknown mechanisms
Embryo		
Cleavage pattern	unequal cleavage lost	change MT/motor localization
Cell fate and cell lineage	cell fate distribution change	heterochrony + pattern change
Larva		
Oral field	reduced degree of differentiation	reduced *gsc* expression time/amount
Larval arms	lost	loss of gene expression regulation
Larval skeleton	reduced	regulation by *otp* lost
Larval gut	lost	heterochrony to late expression
Larval ectoderm	novel structure	fusion of gene expression modules
Expression of genes in ectoderm	unique combinations	co-option/gene regulation changes/gene loss to pseudogene
Adult rudiment		
Development of coelom	early relative to gastrulation	heterochrony
Vestibular ectoderm	early relative to gastrulation	heterochrony
Development of rudiment	early relative to gastrulation	heterochrony
Adult skeleton	accelerated	heterochrony

Nodal is expressed again, this time on the right side of the embryo establishing the L-R axis by repression of left-side differentiation on the right side of the embryo (Duboc et al 2005). Major aspects of the evolution of the *H. erythrogramma* larval form hinge on the control of the timing of axis formation. The most dramatic heterochronic shift is the switch of axis determination to maternal control. This initial pre-formation of axial determination in the egg may allow the very fast development observed in *H. erythrogramma*. This may be a common feature in the switch to direct development; substantial increase in egg size has occurred in the many independent occurrences of direct development in echinoderms (Wray 1996).

A second step affecting axis formation is involved in the loss of differentiation of an overt ventral (oral) ectodermal face in *H. erythrogramma*. The expression of the ventral (oral) end of the A-V axis in indirect developers requires the action of the *goosecoid* gene (Angerer et al 2001). The aborted development of oral ectoderm can be caused experimentally in sea urchins by reduction of expression of a major transcription factor, gsc, in the oral ectodermal domain of the early larva. In *H. erythrogramma*, expression changes in *gsc* are involved the major modifications of ventral ectoderm development leading to the non-feeding larval form (Wilson et al 2005a,b). In this case, the change in timing is a shortened duration of *gsc* expression, resulting in a weakly bounded undifferentiated oral region and loss of the larval mouth. Experimental over-expression of *gsc* mRNA in *H. erythrogramma* results in an enhanced oral ectoderm. The evolutionary reduction in *gsc* expression in presumptive oral ectoderm of *H. erythrogramma* represents a major genic change underlying the shift in larval morphology. This is consistent with the extended period of expression of gsc in *He × Ht* hybrid embryos (Fig. 1C, D), which possess a well-developed oral ectoderm (Wilson et al 2005a,b).

Subsequent to the formation of a reduced D-V axis in *H. erythrogramma*, the larval L-R axis forms precociously relative to indirect developers. In *H. erythrogramma* the L-R axis becomes molecularly and morphologically manifest at the end of gastrulation roughly 24 h after fertilization. In the pluteus larva, L-R axial structures form a few weeks after fertilization (Ferkowicz & Raff 2001). This heterochronic change involves the same signalling system as in the axes of indirect developers, but operating in different developmental stages and in different regulatory environments (Ferkowicz & Raff 2001, Wilson et al 2005a,b, Snoke & Raff unpublished). The developmental timing change involves gene expression heterochronies that initiate the early second expression domain of the Nodal pathway into early larval development. We have found that the Nodal signalling pathway operates in the execution of both these axes similarly in *H. erythrogramma*, but the events are heterochonically shifted in time (Minsuk & Raff 2005, Snoke & Raff, unpublished data).

Together these heterochronic events in regulation of axial pathways drastically speed development, nearly abolish oral side differentiation and morphogenesis, and accelerate differentiation and morphogenesis of structures along the L-R axis that begin the development of the juvenile adult rudiment. In consequence, *H. erythrogramma* metamorphoses in three to four days as opposed to about a month to six weeks required for indirect-developing species. But, it is not only a global heterochrony. The developmental sequence itself is strongly affected.

Not all significant changes are produced through heterochronic changes in the operation of conserved regulatory systems; reorganization has also occurred. Thus, the larval ectoderm of *H. erythrogramma* is not divided into distinct oral and aboral ectoderms as is the pluteus ectoderm. In the pluteus, oral and aboral ectoderm have distinct patterns of gene expression. The *H. erythrogramma* larval ectoderm is unitary and exhibits a unique mix of gene expression patterns consistent with components of both the separate oral and aboral ectoderms of the pluteus. Given the shared upstream pathways, evolution of the *H. erythrogramma* larval ectoderm apparently involved regulatory site changes in downstream genes and novel responses to upstream regulation (Love & Raff 2006). Finally, there also are reductions in developmental features and gene expression. A number of structures found in the pluteus, such as the arms, gut, and specialized aboral ectoderm have been lost in *H. erythrogramma*. Genes specifically expressed in these structures are either no longer expressed in the larva, are expressed but in other patterns, or have become pseudogenes (Love & Raff 2006 unpublished data).

Conclusions

Table 1 summarizes the range of changes in distinct phases of development that have evolved in *H. erythrogramma* since its divergence from *H. tuberculata*. Egg size and lipid content arise from a longer duration of oogenesis as well as a switch to a new phase of greatly reduced yolk addition and accelerated lipid addition (Byrne et al 1999). Lipid content also changes qualitatively (Villinski et al 2002, 2004). The changes in embryonic form that result from cell cleavage and fates and in early onset, and accelerated development of adult features in the larva represent several poorly understood process changes in addition to heterochronies. Other processes lead to the dramatic remodelling of larval morphology. Losses or reductions of larval features stem from heterochronies as well as significant changes in gene regulation (e.g. loss of regulation of larval skeleton by the transcription factor Otp) (Zhou et al 2003). Then there are interesting gains. In a few cases, we have detected high levels of expression of genes not expressed in the pluteus but co-opted for ectodermal expression in *H. erythrogramma* (summarized in Love & Raff 2006). In addition, the novel morphology of the *H. erythrogramma* larva also includes a complex fusion of expression of downstream genes found in the distinct ectoderms of the pluteus, but expressed together in the unitary larval ectoderm of *H. erythrogramma*. As axial regulation involves the same upstream systems in the pluteus and in *H. erythrogramma*, changes in the response of these downstream genes is indicated. The result is a tissue not strictly homologous to either ectoderm of the pluteus. Thus, the apparent developmental simplification of *H. erythrogramma* conceals a number of changes that have produced novel features as well as losses.

The upshot seems to be that evolutionary transformations that would appear to 'macroevolutionary' in size can result from changes in microevolutionary events associated with speciation or following speciation. Some changes are likely to have small genetic costs, but facilitate further significant changes. The gain of a large egg, obviating the need for feeding structures in larvae has been suggested to be key to the cascade of further changes (Wray 1996). Just what selection operates on in the steps of larval evolution is not yet known.

Development of complex multicellular organisms involves modules of various kinds. These allow for the appearance of novel features through two kinds of 'tinkering.' The first is by creation of new features by novel recombination of existing modular elements such as gene expression or developmental heterochronies, changes in sites of gene expression, or changes in signaling between cells or tissues. The second mode of tinkering is by creation of new modules through duplication and divergence of pre-existing ones, or through module fusion, e.g. in genes or developmental fields, to provide additional modules that can enter the pool of entities available for further tinkering. *H. erythrogramma* embryos illustrate both of these kinds of processes. A dramatic reorganization of existing modules is seen to have occurred in moving embryonic axis formation from embryogenesis to oogenesis. A new module has arisen in the formation of the larval ectoderm, a tissue that is not clearly homologous to its evolutionary precursors. Genic changes that underlie major evolutionary events must in the main be the same kind of changes that occur in microevolutionary processes within species. The change in 'scale' from micro- to macroevolution is a phenotypic result that can occur quickly as a consequence of changes that affect modules, coupled with the appropriate selection. Thus, we do not need to invoke macroevolution as necessarily involving rare 'different' kinds of events—although such events should also sometimes occur.

Acknowledgements

We thank our colleagues at Indiana University and the University of Sydney, and the NSF for support of our research.

References

Angerer LM, Oleksyn DW, Levine AM, Li X, Klein WH, Angerer RC 2001 Sea urchin goosecoid function links fate specification along the animal-vegetal and oral-aboral embryonic axes. Development 128:4393–4404

Angerer LM, Angerer RC 2003 Patterning the sea urchin embryo: gene regulatory networks, signaling pathways, and cellular interactions. Curr Topics Dev Biol 53:159–198

Bininda-Emonds ORP, Jeffery JE, Richardson MK 2003 Inverting the hourglass: quantitative evidence against the phylotypic stage in vertebrate development. Proc Biol Sci 270:341–346

Byrne M 2006 Life history diversity and the Asterinidae. Integr Comp Biol 46:243–254

Byrne M, Villinski JT, Cisternas P, Popodi E, Raff RA 1999 Maternal factors and the evolution of developmental mode: Evolution of oogenesis in *Heliocidaris erythrogramma*. Dev Genes Evol 209:275–283

Collin R 2004 Phylogenetic effects, the loss of complex characters, and the evolution of development in calyptraeid gastropods. Evolution 58:1488–1502

Duboc V, Röttinger E, Besnardeau L, Lepage T 2004 Nodal and BMP2/4 signaling organizes the oral-aboral axis of the sea urchin embryo. Dev Cell 6:397–410

Duboc V, Röttinger E, Lapraz F, Besnardeau L, Lepage T 2005 Left-right asymmetry in the sea urchin embryo is regulated by Nodal signaling on the right side. Dev Cell 9:147–158

Ferkowicz MJ, Raff RA 2001 Wnt gene expression in sea urchin development: heterochronies associated with the evolution of developmental mode. Evol Dev 3:24–33

Flowers VL, Courteau GR, Poustka AJ, Weng W, Venuti JM 2004 Nodal/activin signaling establishes oral-aboral polarity in the early sea urchin embryo. Dev Dyn 231:727–740

Gibson GD, Gibson AJF 2004 Heterochrony and the evolution of poecilogony: generation of larval diversity. Evolution 58:2704–2717

Hart MW, Byrne M, Smith MJ 1997 Molecular phylogenetic analysis of life-history evolution in asterinid starfish. Evolution 51:1848–1861

Henry JJ, Raff RA 1990 The dorsoventral axis is specified prior to first cleavage in the direct developing sea urchin *Heliocidaris erythrogramma*. Development 110:875–884

Jeffery CH, Emlet RB 2003 Macroevolutionary consequences of developmental mode in temnopleurid echinoids from the Tertiary of southern Australia. Evolution 57:1031–1048

Jeffery WR, Swalla BJ, Ewing N, Kusakabe T 1999 Evolution of the ascidian anural larva: evidence from embryos and molecules. Mol Biol Evol 16:646–654

Jeffery CH, Emlet RB, Littlewood DT 2003 Phylogeny and evolution of developmental mode in temnopleurid echinoids. Mol Phylogenet Evol 28:99–118

Kauffman JS, Raff RA 2003 Patterning mechanisms in the evolution of derived developmental life histories: the role of Wnt signaling in axis formation of the direct-developing sea urchin *Heliocidaris erythrogramma*. Dev Genes Evol 213:612–624

Love A, Raff RA 2006 Larval ectoderm, organizational homology, and the origins of evolutionary novelty. J Exp Zoolog B Mol Dev Evol 306:18–34

Love AC, Andrews ME, Raff RA 2006 Gene expression patterns in a novel animal appendage: The sea urchin pluteus arm. Evol Dev 9:51–68

Minsuk S, Raff RA 2005 Co-option of an oral-aboral patterning mechanism to control left-right differentiation: the direct-developing sea urchin *Heliocidaris erythrogramma* is sinistralized, not ventralized, by NiCl2. Evol Dev 7:289–300

Nielsen MG, Wilson KA, Raff EC, Raff RA 2000 Novel gene expression patterns in hybrid embryos between species with different modes of development. Evol Dev 2:133–144

Raff RA 1996 The Shape of Life. University of Chicago Press, Chicago

Raff RA, Herlands L, Morris VB, Healy J 1990 Evolutionary modification of echinoid sperm correlates with developmental mode. Dev Growth Differ 32:283–291

Raff EC, Popodi EM, Sly BJ, Turner FR, Villinski JT, Raff RA 1999 A novel ontogenetic pathway in hybrid embryos between species with different modes of development. Development 126:1937–1945

Raff EC, Popodi EM, Kauffman J et al 2003 Regulatory punctuated equilibrium and convergence in the evolution of developmental pathways in direct-developing sea urchins. Evol Dev 5:478–493

Shapiro MD, Marks ME, Peichel CL et al 2004 Genetic and developmental basis of evolutionary pelvic reduction in threespine sticklebacks. Nature 428:717–723

Smith, MJ, Boom JDG, Raff RA 1990 Single-copy DNA distance between two congeneric sea urchin species exhibiting radically different modes of development. Mol Biol Evol 7:315–326

Villinski JC, Hayes JM, Villinski JT, Brassell SC, Raff RA 2004 Carbon-isotopic shifts associated with heterotrophy and biosynthetic pathways in direct- and indirect-developing sea urchins. Mar Ecol Prog Ser 275:139–151

Villinski JT, Villinski JC, Raff RA 2002 Convergence in maternal provisioning strategy during developmental evolution of sea urchins. Evolution 56:1764–1775

Wilson K, Andrews MA, Raff RA 2005a Dissociation of expression patterns of homeodomain transcription factors in the evolution of developmental mode in the sea urchins *Heliocidaris tuberculata* and *H. erythrogramma*. Evol Dev 7:401–415

Wilson K, Andrews MA, Turner FR, Raff RA 2005b Major regulatory factors in the evolution of development: the roles of goosecoid and Msx in the evolution of the direct-developing sea urchin *Heliocidaris erythrogramma*. Evol Dev 7:416–428

Wray GA 1996 Parallel evolution of nonfeeding larvae in echinoids. Syst Biol 45:308–322

Wray GA, Raff RA 1990 Novel origins of lineage founder cells in the direct-developing sea urchin *Heliocidaris erythrogramma*. Dev Biol 141:41–54

Zhou N, Wilson KA, Andrews ME, Kauffman JS, Raff RA 2003 Evolution of OTP-independent larval skeleton patterning in the direct-developing sea urchin, *Heliocidaris erythrogramma*. J Exp Zoolog B Mol Dev Evol 300B:58–71

Zigler KS, Raff EC, Popodi E, Raff RA, Lessios HE 2003 Adaptive evolution of bindin in the genus *Heliocidaris* is correlated with the shift to direct development. Evolution 57:2293–2302

DISCUSSION

Bell: You showed a cladogram that depicts the evolution of a 'shmoo' (a larval echinoderm that does not feed and therefore is fairly featureless) in three separate places. I hoped that you'd be telling us about the similarities and dissimilarities of the genetics of the formation of that shmoo in different groups. Do you know any more about this than you have told us?

R Raff: If you make a hybrid between the independently evolved shmoo forms you get another shmoo and it metamorphoses into a sea urchin. This is suggestive that we are looking at similar genetic machinery in operation. We don't have any data on these other species. No one has worked on other direct developers to any great extent.

Lieberman: What is the fitness of these hybrids? Do you have any sense that they are suffering fitness problems?

R Raff: Their fitness is probably zero. This sort of hybridization doesn't occur in nature. We have to do things to make them fertilize. They metamorphose and we get little sea urchins, but most of them have behavioural defects. Perhaps some could be raised up, but we would have to be there all year to raise these in large enough numbers to get a sample that could be fertile.

Wagner: What was the evidence for large-effect genes as part of this transformation?

R Raff: This comes from looking at the transcription factor goosecoid. Its downregulation and shorter expression time seems to be involved in reducing the differentiation of the oral site in *H. erythrogramma*.

Wagner: This doesn't exclude that the actual evolutionary transformation was through small effect alleles.

R Raff: That's right. The other evidence we have that large effects are taking place comes from the ectoderm of the pluteus. The pluteus has two kinds: oral and aboral ectoderm. In *H. erythrogramma*, there is only one kind of ectoderm, and it is a combination of the two in terms of its genetic markers and cells types. Thus, there is a lot of downstream gene expression change, even though the upstream events seem to be pretty conserved.

Brakefield: You implied that the life history evolution is crucial in the whole story. Do you have any thoughts about the genetic basis of the difference in egg size, and do you have any indication about the number of genes involved there? Are the hybrids intermediate in egg size?

R Raff: No, because the cross is viable in only one direction—with *H. erythrogramma* eggs. We only have two lines of evidence on evolution of large egg size. With Maria Byrne of the University of Sydney, we looked at oogenesis in the big egg species. Until the large egg reaches about 90 μm it is like the oocyte of the smaller egg of *H. tuberculata*, then it stops vitellogenesis and goes into a whole new phase of addition of lipids and non-yolk to make the full-sized egg, which is 100 times the volume of the *H. tuberculata* egg. This is a novel phenomenon that is added on at the end of oogenesis. We also know that there is a lot of co-evolution between the eggs and sperm. In every case we get these big eggs we get large sperm with heads like bananas. These species with big eggs have larger genomes than their related indirect developers. This is as close to genetics that we have got.

Wilkins: You are looking at the evolutionary end result of a million or more generations, which is a long time. You have noted a lot of changes occurring, but we don't know the sequence. Yet there must have been a sequence, and some of the first changes may have produced selective pressures for subsequent changes. It is impossible to work out what this sequence is just from comparing these two species, but can one use other direct-developing sea urchins which don't have as complete a direct development to begin to work out what this sequence of events might have been?

R Raff: A theoretical paper was published a number of years ago which pointed out that there were two choices (Vance 1973): an infinite number of small eggs, or a small number of really big eggs. Of the 1000 or so sea urchin species, 80% of them produce small eggs and a feeding larva, about 20% are direct developers, and only two species are in the middle (they make big eggs but make a pluteus that feeds). The species that I showed (*Clypeaster rosaceus*) is the only one of this latter category that can easily be studied. This is an important data point, because they are capable of feeding and making a nice pluteus early on, but there is an acceleration in making left coelom and a speed-up to metamorphosis. This suggests that it is not just a matter of trophic evolution, but instead there are selective pathways, one towards having larger eggs in order to omit the feeding, and the other an early onset and acceleration in production of the adult rudiment. Whatever the actual events are, there is probably some advantage to shift not only to non-feeding, but also to higher developmental gear. Probably the first step is to turn on early production the left coelom, the key to rapid development of the rudiment of the adult.

Wilkins: Perhaps this is the initial step that then creates selective pressure for further genetic changes.

R Raff: I have no doubt about that.

Budd: In terms of tinkering or *bricolage*, these initial steps weren't for the embryos to become non-feeding. There must have been some other kind of selective advantage for these initial changes, in order to maintain some selective functional continuity throughout this process.

R Raff: There seems to be an economic connection. In Antarctica, where there are 20 species of sea urchin, 19 are direct developers. In the deep sea, most sea urchins are direct developers. It probably has to so with things that affect larval success. In some cases it may have to do with poorly predictable plankton, or disperal phenomena. It is very much related to population phenomena.

Budd: It is fascinating that there has been a repeated evolutionary transformation from a pluteus to a shmoo, and yet the intermediate stages are highly unstable in the evolutionary sense.

Carroll: You mentioned that it was a pity that there were no hexapod vertebrates. There were: the agnothodians in the Devonian.

R Raff: True, but I was talking about tetrapods.

Carroll: There was a time in the history of vertebrates when there was an enormous amount of flexibility in the number of fin spines and a possible non-homology from the forelimb and the hindlimb. At that particular time in evolutionary history there was the potential for an enormous flexibility that was then pared down.

R Raff: I agree: I was really only dealing with the tetrapod radiation. It is not clear that you couldn't duplicate limb buds, which could produce more sets of limb pairs, except that it may involve such deeply ingrained Hox patterning that it might not be viable.

Wagner: I would rather worry about the functional integration of another pair of limbs in CNS function, and all those contingencies that need to be there in order to take advantage of another pair of limbs.

R Raff: If you could do it by patterning, it might well be possible for innervation to occur.

Duboule: You can get supernumerary limbs. It is a teratology observed in the human population. Interestingly, if you get them at the level of the abdomen, they are always ventral and they look like a hind limb. If you get them at the level of the shoulder they always look like an arm. It is not viable for the reasons that you mentioned: you need such a reorganization of the trunk patterning system.

Lieberman: You could take the data you mentioned and make the reverse argument. You could argue that although you get these mutations of large effect, you have also shown that they are not all that viable. The evolution that would have been necessary for the multiple evolution of the direct developing forms to occur would have involved selection acting on different parts of the pathway.

R Raff: First of all, the fact that you don't find many things lying between indirect and direct development may be a matter of the rate at which it takes place. It could be that if it is unstable, they simply go back to the stable indirect-developing end point. There is quite a range of egg sizes allowing either. The other possibility is that they fall into the path funnelling down towards the shmoo. This decision might be made fairly rapidly, so there may not be many such species out there at any particular time, even though this is happening more or less continuously and has taken place in many clades.

Stern: It seems to me that the integration argument is a bit of a red herring. Viability of an organ isn't required for it to be selected. For example, there can be sexual selection for some weird organ. It could be an extra set of limbs that isn't connected to the CNS. Perhaps later these could be integrated by having selection for them to work. This is rather an extreme example, but there are ways for organs to be selected in populations even if they don't appear viable.

Duboule: I meant that the organism is not viable. If you want an extra limb bud at this position you need factors coming from internal organs, and so on. There are, however, cases where there is a sort of bifid arm. There was a recent surgical procedure on a girl in China where she had a perfect additional supernumerary arm, which was by itself not viable. This is very rare though.

Bell: I want to make a comment about the relationship between morphological series, which have been alluded to here on a number of occasions, to try to explain evolutionary transitions. Morphological series represent standing variation within a population, a species or some higher taxon. We now have a number of cases of contemporary evolution in threespine stickleback (*Gasterosteus aculeatus*) populations (e.g. Bell et al 2004). There are two things we see. One is that there are phenotypes seen in real evolutionary transitions that aren't seen in the morphological series within populations. There are transient polymorphisms that include phenotypes that are absent from morphological series. This suggests that it is dangerous to infer evolutionary transitions from morphological series of existing populations, species or higher taxa. The second is that evolutionary transitions can occur incredibly fast. In our organism these are major morphological differences. Much of the diversification that exists in stickleback could have evolved in 10–20 generations. People marvel at how much diversity there is in post-glacial populations: it may have happened in 10 generations. I would caution people against making inferences from morphological series within well established species. The intermediate species aren't on their way to achieving the derived state. They may be in a stable state of their own, which has nothing to do with evolutionary transition.

Morriss-Kay: What you have described is a major change in the way development proceeds, yet almost no difference in the outcome, i.e. evolutionary pressures have created very different larval morphologies which nevertheless form adult organisms that are essentially the same. I can think of two other relevant examples. The most obvious one is formation of the neural tube in different groups of vertebrates: a hollow spinal cord can form either by folding up a flat sheet into a tube or by canalisation of a solid structure. The contrasting morphogenetic processes that generate similar eye morphologies in cephalapods and vertebrates is even more striking. I'm still trying to get to grips with the idea of what we mean by tinkering, and this idea that you can call it tinkering when you get such a major change early in development and yet end up with something that is the same. There seem to me to be evolutionary mechanisms that enable a particular outcome because the adult organism has a good evolutionary niche to survive, even though the pressures on the larva force it to get to that outcome through a different route.

R Raff: This was noted by some of the embryologists at Woods Hole in the 1890s. There were organisms that make an egg and sperm, fertilization occurs and all the tissue layers form. Some of these organisms are also capable of budding, producing a young animal with the same layers of tissue but without going through embryonic development. It happens. In the case of these larval forms, there is a real bifurcation in developmental patterns. Larvae have evolved to the point where much of development is of the vehicle that carries the cells that will differentiate to become the juvenile. Much of the evolutionary change we observe in larval evolution involves those larval parts and does not affect the cells and cell interactions that are going to produce the juvenile *per se*. The larva is almost a separate creature in its own right, with its own morphology and ecological adaptations.

Lieberman: In a way you proposed that many of the mechanisms by which these changes occur is via alterations in timing. There is a heterochrony which you were surprised about. But in a sense heterochrony is a perfect kind of tinkering. The machinery is already there to do it. In a sense when Gould and Lewington and others started talking about contingency and heterochrony they were raising ideas of tinkering.

R Raff: The thing about the heterochronies seen in these derived larval forms is that the modules are doing the same thing but they are doing it in a setting that is different. This suggests a fair amount of autonomy. Indeed, the vestibular ectoderm that forms the ectoderm in the adult is highly autonomous in its behaviour in early development. In a way, this makes possible the heterochronies. The machinery already exists and there is a fair amount of autonomy in the modules.

Hall: Could you tell us about this in relation to the speeding up in the left coelom? Can you trace this back earlier in development, showing that it involves things such as germ layer cells being specified into different regions? Can you get some sort of sense as to how labile early development is?

R Raff: We've found a surprising evolutionary lability in early development, including cleavage patterns, cell lineages, and heterochronies. In the case of the left coelom, the mechanism seems to be a difference in how the cells arise. That is, in an indirect developer a bunch of cells that form from the small micromere daughters form small pouches on either side of the gut. They sit there for a long time. Eventually, they start dividing and produce a lot of cells that give rise to those long structures. In *H. erythrogramma*, it doesn't look as if there is much cell division taking place. It looks as if the cells that are moving in with the archenteron are then changing from an archenteron specification to a coelomic one, and cells are just moving into that fate. The cells to form the pouches are not arising by cell division in this case.

Budd: It is striking that we are dealing with successive losses of structures. Of course, there are important developmental implications for this. Nevertheless, we are losing certain types of feeding and locomotion structures over and over again.

R Raff: We are gaining a larval ectoderm, which is a completely different ectoderm.

Budd: The whole point of being a larva is to fulfil certain functions that will allow your egg to get through to being an adult. Once you start generating redundancies in some of those mechanisms, then you can free up and lose all the stuff you used to use before. Once you have developed a large egg and perhaps shortened your development a bit, then you can have enough resources to do away with the need to feed. It is not surprising, therefore, that there could then be rapid loss of the structures needed for feeding. What would be interesting would be to see whether the opposite could occur. Do you ever see repeated evolution of novel structures in order to carry out a function?

R Raff: We thought about this years ago because there are some direct developers that still have arms on them. They don't have a gut but they still have some remnant arms. Are these shmoos that are going back, or plutei that are going forward? We did some calculations on how long it takes some genes to die (Marshall et al 1994). In the course of a couple of million years, it could be either one. There might be a fair amount of oscillation going on that we are not detecting in processes like that. It seems rare for a non-feeding shmoo to reverse itself and give arise to a novel feeding larva.

Weiss: This raises a question that people have been talking about in human evolution, especially since the availability of the chimpanzee and macaque genomes.

This is the idea that humans are very different from chimps despite the short evolutionary time because of gene loss. That suggests evolution that was more rapid than the assembly of something new by gene duplication and subsequent tuning by selection. Maynard Olson has said this is most likely to be achieved by gene loss. As I understand it, the data from comparisons of the chimp and human genomes show evidence for gene gain and major segmental duplication (and also some gene loss) in both lineages. Part of the problem is that we don't know the genetic basis of anything complicated enough to see whether it has been assembled by rapid accumulation of lots of small changes, or something simple that has been lost in terms of organization, or a simple gain?

Wilkins: Don't you question, however, the basic assumptions that chimps are very different from humans and there hasn't been enough time for those extensive genetic changes?

Weiss: King & Wilson (1975) hypothesized that it was small regulatory changes. People view this as a prescient paper because at the time *cis*-regulation was only known in some experimental systems. Their idea was that it was a small number of regulatory changes. Gene duplication was discovered after this. Recently, people have begun looking for gene loss and duplication.

Wagner: We probably underestimate the molecular mechanisms that can lead to developmental changes. 60% of the transcripts from the human genome are small non-coding RNAs. Wherever we look there are new regulatory functions at all kinds of levels of gene expression. The only area where we have enough comparative data is with microRNAs. There seems to be an incredible rate of innovation in terms of microRNAs, even among closely related *Drosophila* species. This could be a huge source of molecular genetic variation leading to differential regulation.

Weiss: This hints at a big mix of quantitative regulatory factors that can be moulded gradually, rather than some major jump.

Bell: Five to eight million years is a lot of time for selection to act. The availability of genetic variation is a separate issue. There is ample time for selection to make chimps and humans from one common ancestor.

Weiss: The debate on these points often becomes arcane and vicious among anthropologists! When do you hypothesize that a creature had the modern human mental abilities? Has this only been a couple of hundred thousand years? Back 9 million years, it becomes less remarkable.

R Raff: Why would you put it in terms of these diametric changes? Humans and chimps share many cognitive characteristics. Along the way there will have been a mosaic evolution of the brain just as there is a mosaic evolution of larvae.

Lieberman: You would be surprised how controversial a comment this would be at a human evolution conference.

Hall: To take natural selection back to sea urchins would be a way of examining mosaic evolution. One of the advantages of having two life history stages is that you can have independent selection of the adult and larva. Here you have a situation where the larva has the adult rudiment in it. How much of the selection on the larva is actually for larval characteristics for the larva to survive, and how much is the selection for the larva as a vehicle to get the adult rudiment through to the adult stage?

R Raff: We don't know enough. We don't even know the full range of developmental phenomena. A few years ago there was a description of a brittle star that also makes a pluteus. In this particular species the pluteus grows, makes a rudiment, metamorphoses and then the rudiment walks off. The remaining larval part reassembles a bunch of cells, gastrulates again and does it all over. You tell me what is being selected for!

Coates: Returning to this contrast between life histories, is there any evidence of differing evolutionary rates between the two strategies? If you examine echinoderm clades, is there evidence to suggest that the direct developers show evidence of greater evolutionary divergence (assuming that the dispersal of direct developers is much more limited)?

R Raff: Charlotte Jeffery and her co-workers looked at this in another clade (Jeffery & Emlet 2003, Jeffery et al 2003). Among the fossils they found lineages that were direct developers. These have short durations so there is repeated convergent evolution going on from stocks that likely had pluteus larvae. Among the living direct developers we wanted to look at how fast the evolution was. We know that *H. erythrogramma* bindin, a protein involved in fertilization, evolves much more rapidly than it does in *H. tuberculata*. What about development? We used a hybrid strategy to ask if *H. erythrogramma* developmental mechanisms have evolved more rapidly because we know what the hybrid phenotypes are when we cross *H. erthyrogramma* and *H. tuberculata*. What happens if you take an indirect-developing sea urchin from another family and cross it to *H. erythrogramma*? We took one from a family which is about 30 million years away, and made hybrids with these. They give the same syndrome as the closer ones do, even though we are eight times further away in time. The hybrids look pretty much the same as *erythrogramma* times *tuberculata*. This would suggest that there is more evolution on the *H. erythrogramma* side than on those species that develop via a pluteus.

Budd: In terms of loss and gain issues, there have been a couple of studies (e.g. Goloboff 1997) trying to show the balance of loss versus gain in different lineages. There was a study on stick insects (Whiting et al 2003) that tried to claim wings had re-evolved. But not everyone is entirely happy with that phylogenetic reconstruction. It is hard to show that there are situations with a bunch of regulatory genes sitting around which then get redeployed to effectively resurrect the old structure that was lost in the lineage.

Wagner: We have done some work on digit loss and evolution in a genus of lizards in South America. In this case we showed rigorously that digits re-evolved using comparative methods and a well supported phylogeny. In this context we looked at other examples. We found a handful of cases where there is statistical support for re-evolution of complex structures. For establishing this you need six or seven outgroup lineages that don't all have the structure.

Stern: It is not necessarily re-evolution, but the evolution of endosymbiosis has been very common. Many bacteria have become integrated into cells of many different species independently.

Lieberman: What about teeth? Is there is significant evidence for Dollo's law being violated by teeth?

Jernvall: Modern lynx frequently have two sets of lower molars. In the fossil record felines have lacked the second molar since the Miocene. Lynx is a fairly accepted example of where a lost tooth has reappeared during evolution. This could be a perfect example of tinkering.

Wagner: In terms of digits re-evolution, a nice side observation in our study is that according to our phylogeny, in the species that have re-evolved digits, these digits are morphologically different from those that are ancestral. All the re-evolved digits have only two phalanges. There seems to be a redeployment of the digit developmental program, but a lot of morphological detail has been lost.

R Raff: There are life history reversals too, best studied in salamanders. There seems to be a turning on and off of neotony. There the nodes are very short.

Olsson: There is a controversy in salamanders. The question is whether in one group of lungless salamanders the larva has re-evolved or not. In the phylogeny all the basal species have direct development, and further out they have a larva. There could be multiple losses.

Wagner: In snakes there is also evidence that ovipary was re-evolved.

Hall: I want to comment on a new book by John Avise (Avise 2006). He has taken numerous molecular phylogenies and mapped onto them the morphological, developmental, physiological or behavioural changes, with four or five dozen examples right across the plant and animal kingdom. If you accept the molecular phylogenies as being an accurate ones, in each of these cases you can look and see whether these characters have re-evolved or not.

Coates: My impression on looking at the vertebrate record for patterns of anatomical loss and re-evolution, is that when appendages and other structures are lost, it is their size that goes first: the anatomical pattern is maintained. It is rare to see pattern depart first, and this suggests that the developmental kit is robust and persists.

Hall: Yes, limb buds tend to be formed even if the whole limb has been lost in limbless forms.

References

Avise JC 2006 Evolutionary pathways in nature: a phylogenetic approach. Cambridge University Press, New York

Bell MA, Aguirre WE, Buck NJ 2004 Twelve years of contemporary armor evolution in a threespine stickleback population. Evolution 58:814–824

Goloboff PA 1997 Self-weighted optimization: tree searches and character state reconstructions under implied transformation costs. Cladistics 13:225–245

Jeffery CH, Emlet RB 2003 Macroevolutionary consequences of developmental mode in temnopleurid echinoids from the Tertiary of southern Australia. Evolution 57:1031–1048

Jeffery CH, Emlet RB, Littlewood DT 2003 Phylogeny and evolution of developmental mode in temnopleurid echinoids. Mol Phylogenet Evol 28:99–118

King MC, Wilson AC 1975 Evolution at two levels in humans and chimpanzees. Science 188:107–116

Marshall CR, Raff EC, Raff RA 1994 Dollo's Law and the death and resurrection of genes. Proc Natl Acad Sci USA 91:12283–12287

Vance RR 1973 On reproductive strategies in marine benthic invertebrates. Am Nat 107:339–352

Whiting MF, Bradler S, Maxwell T 2003 Loss and recovery of wings in tick insects. Nature 421:264–267

The relationship between development and evolution through heritable variation

James M. Cheverud

Department of Anatomy & Neurobiology, Washington University School of Medicine, 660 S. Euclid Avenue, St. Louis, MO 63110, USA

Abstract. Darwin's theory of evolution by natural selection states that evolution occurs through the natural selection of heritable variation. Development plays the key physiological role connecting the heritable genotypes, passed from one generation to the next, to the phenotypes that are made available for selection. While at times the developmental variations underlying a selected trait may be neutral with respect to selection, it is through its effects on heritable variation that developmental tinkering affects evolution. We can gain a deeper understanding of the evolutionary process by considering the role of development in structuring variation and, through its effects on variation, structuring evolution. Both evolutionary theory and empirical studies show that features that interact in development tend to be inherited together and, hence, to evolve together. Gene mapping studies show that this modular inheritance pattern is due to modular pleiotropic gene effects, individual genes affecting a single modular unit, and that there is heritable variation in the range of features encompassed by these modules. We hypothesize that modular pleiotropic patterns are sculpted by natural selection so that functionally-and developmentally-related traits are affected by module-specific genes.

2007 Tinkering: the microevolution of development. Wiley, Chichester (Novartis Foundation Symposium 284) p 55–70

The developmental process provides the physiological relationship between genotype and phenotype encompassing the means by which genetic and environmental factors affect the developing phenotype. However, evolution is not directly affected by this physiological relationship. Instead, evolution depends on the statistical relationship of genotype to phenotype. Even so, the physiological relationship between genotype and phenotype structures this statistical relationship. To relate development to evolution, we need to relate the physiological bases of development to the statistical relationships between genotype and phenotype.

Over a decade ago, Atchley & Hall (1991) proposed a model relating developmental processes to the statistical relationship between genotype and phenotype

Atchley-Hall Model

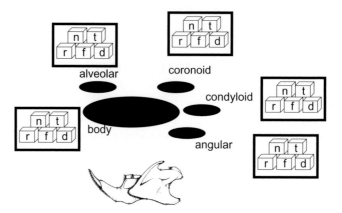

FIG. 1. The Atchley-Hall Model for the developmental basis of variation in morphological size and shape. The mandible is composed of five cellular condensations (coronoid process, angular process, condyloid process, body, alveolus). The final size of each condensation depends on five developmental parameters: n (number of cells in condensation), t (time of condensation initiation), r (rate of cell division), f (fraction of cells mitotically active), d (rate of cell death).

(see Fig. 1). In this model, the size of a developing organ at some time 't' (n_t) was considered as a function of the number of cells (n_0) and the time (t_0) of the initial condensation, the rate of division of individual cells (r), the fraction of cells mitotically active (f), and the rate of programmed cell death (d),

$$n_t = n_0(1 + (t - t_0)rf - (t - t_0)d).$$

The genetic variance of an organ's size and shape would then be a function of the genetic variances and covariances among these developmental parameters, thereby relating variation in the developmental parameters to the variation in the phenotype.

Developmental homoplasy

Using this model, Atchley et al (1997) noted that any given phenotypic value (n_t) could be obtained through a variety of combinations of the developmental parameters. Hence, specific developmental parameters may be neutral in relation to selection on the final phenotype. They referred to this phenomenon as developmental homoplasy, many different developmental mechanisms leading to the same phenotypic end result. In this situation, the developmental underpinnings of the evolved trait depend solely on the genetic variances of the developmental

parameters and their genetic covariances with the selected trait, the end phenotype. While it is certainly possible that the developmental parameters themselves
could be subject to selection, in the situation where selection is on the end phenotype, the precise developmental mechanisms underlying evolutionary change are
neutral, and hence not specifically important for an understanding of the evolutionary process.

Two examples of developmental homoplasy serve to illustrate the general
principle. Oxnard (1976, 1984) studied morphological variation in relation to locomotion in prosimian primates. He contrasted species from genera exhibiting generalized quadrapedal locomotion (e.g. *Lemur*) with vertical clinging and leaping
species (e.g. *Indri, Avahi, Propithecus, Galago, Tarsius*). The leaping species have relatively long legs which are commonly interpreted as an adaptation to leaping forms
of locomotion because the force exerted during leaping is proportional to the
change in leg length during lift off. Longer legs lead to more powerful leaps.
However, a multivariate analysis of limb segments in these forms indicated that
while the leaping prosimians all had relatively long legs, they accomplished this
through different developmental mechanisms. Members of the genera *Indri, Avahi*
and *Propithecus* have relatively long femora and metatarsus, the first and last elements in the limb, while *Galago* and *Tarsius* have relatively long middle segments,
tibia and tarsus. It is important to note that both these groups represent multiple,
independent origins of long legs and leaping. Selection for increased leg length in
leaping prosimians resulted in the evolution of different limb elements in different
groups. Which elements increased in length in these groups was likely determined
by which developmental variations happened to be present in their ancestral forms
rather than representing a specific adaptation.

A second example of developmental homoplasy is the result of a selection
experiment for longer tails in mice. Rutledge et al (1974) formed two small initial
populations from a large random-bred mouse population to serve as replicate
selection lines for increased tail length. The experiment succeeded and both
selected lines evolved similarly long tails. However, upon further examination they
found that one line accomplished this by making individual vertebra longer while
the other increased the number of vertebra in the tail. This diversity in developmental response was due to random founder effects involved in the formation of
the original selected lines. One line happened to vary for vertebral length and the
other for vertebral number. Different developmental processes underlay the
common response to selection for longer tail length.

Elements of evolutionary theory

Even though in these examples the mechanisms underlying development are
neutral, they can still have consequences for further evolution. Development may

have a prospective effect on future evolution even when it is neutral in retrospect. How does development intersect with evolutionary processes? To answer this question, we need to consider basic evolutionary theory at the level of first principles.

Darwin's theory of evolution by natural selection states that evolution occurs through the natural selection of heritable variation. Natural selection is due to a correlation between fitness and some phenotype caused by the interaction between phenotype and environment. Heritable variation refers to the heritable differences between individuals. In the early 20th century evolutionary geneticists put Darwin's theory on a clear, logical and mathematical basis (Provine 1976). They showed that the heritable variation in a population was measured by the additive genetic variance. When many traits are considered together, heritable variance takes the form of a square symmetric matrix with additive genetic variances along the diagonal and additive genetic covariances between traits off the diagonal, the genetic variance/covariance matrix (G). The additive genetic variance measures the amount of heritable variance for the trait while the additive genetic covariances measure the degree to which different traits are inherited together. This co-inheritance is due to pleiotropy, where one gene affects several traits, and/or linkage disequilibrium, where two different genes, each affecting a separate trait, tend to be inherited together most often because they lie near each other along the chromosomes. In seeking the interface between development and evolution perhaps the most fruitful enquiry considers the effects of developmental processes in moulding patterns of heritable variation.

Heritable variation is itself an evolving property of a population. Two major factors affect the evolution of heritable variation; the pattern of new variation produced by mutation and the pattern of stabilizing selection on the traits (Lande 1980). While mutation may be random with respect to selection, it is unlikely that all traits mutate at the same rate or mutate independently of each other. Mutation will produce high variance in some traits and lower variance in others. Likewise, a range of traits will be affected by the pleiotropic effects of new mutations while other traits are unaffected. Theoretical predictions for the evolution of the mutation variance/covariance matrix are not available but it seems possible that the effects of mutations are channelled by developmental processes. Stabilizing selection also causes the evolution of heritable variance/covariance patterns. Traits that interact during development or to perform an adult physiological function will be co-selected because the fitness effect of one trait depends on the value of the second trait. This is sometimes referred to as internal stabilizing selection because the selection is on the relationships of the parts rather than on their direct interface with the environment (Reidl 1976). The selection is to 'fit in' with other aspects of the phenotype. For example, the upper and lower jaws work together in mammalian mastication for food preparation. In a certain instance, it may not matter whether the mandible itself is long or short so long as the upper and lower jaws match, preventing over- and underbite. Selection would act strongly against mismatches, long mandible with short maxilla and vice versa, selecting to correlate the inheritance of these two traits.

Morphological integration

The logic of these relationships is commonly considered under the principle of morphological integration (Olson & Miller 1958). Morphological integration refers to the relationships between morphological parts. Olson & Miller (1958) postulated that developmentally and/or functionally related traits would be correlated in their distributions within populations and evolve jointly rather than mosaically. Later quantitative genetic theory supported their arguments by indicating that functionally and developmentally related traits should be inherited together because patterns of stabilizing selection lead to the co-inheritance of related traits (Lande 1980, Cheverud 1982, 1984) and that because of this co-inheritance, functionally and developmentally related traits should evolve together through correlated responses to selection (Lande 1979). The phenotype is integrated, not atomistic.

In a series of studies on the co-inheritance of cranial morphology in non-human primates (Cheverud 1982, 1996a), it was indeed found that functionally and developmentally-related cranial traits are relatively highly genetically correlated. Related traits share about twice the variance shared by unrelated traits. This general pattern is also evident in the phenotypic patterns of cranial integration throughout the primates (Marroig & Cheverud 2001, Ackermann & Cheverud 2004). The primary source of this patterning is a contrast between the braincase and face of the skull. The primate braincase grows early in life, in step with the growing brain that it serves to enclose and protect. In contrast, the face, and especially the masticatory apparatus, grows later, along with the rest of the body. Both genetic and phenotypic correlations are higher within these parts of the skull than between them. They are individuated relative to one another forming separate cranial modules.

Modular or antagonistic pleiotropy

Co-inheritance of developmentally and functionally related traits can occur through two different genetic mechanisms; pleiotropy can have an antagonistic pattern with each locus affecting each trait but with opposite effects on different developmental regions or processes, or it can display a modular pattern, with the effects of subsets of genes being restricted to specific subsets of developmental units (see Fig. 2). These two possibilities reflect different styles of developmental processes

Modular Pleiotropy   Antagonistic Pleiotropy

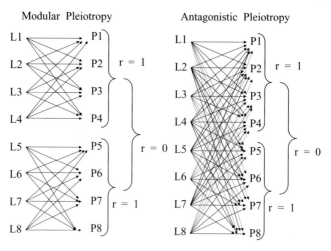

FIG. 2. The two possible pleiotropy patterns responsible for genetic morphological integration; modular pleiotropy and antagonistic pleiotropy. L*i* refers to different gene loci and P*i* refers to different phenotypic traits. In antagonistic pleiotropy each locus affects all traits but has opposite effects on different modules. In modular pleiotropy the effects of loci are restricted to a subset of traits.

(Riska 1986) and have divergent evolutionary consequences. However, both can account for the observed co-inheritance of functionally and developmentally related traits.

In order to distinguish between these two possibilities, it is necessary to examine the effects of individual gene loci on morphological characters. This has become possible over the past decades through quantitative trait locus (QTL) analysis (Lynch & Walsh 1998). In a typical experiment of this kind, two inbred strains are crossed to produce F_1 hybrid animals that are heterozygous at all loci different between the two parent strains. The F_1 animals are then intercrossed to produce a F_2 generation. In the F_2 generation genes will segregate according to Mendel's Laws as will any phenotypes they affect. The effects of specific gene regions on morphology can then be discerned by correlating the segregation of genes with the segregation of phenotypes across the whole genome.

We have performed several QTL experiments where we have measured the pleiotropic effects of genes specifically to determine whether modular or antagonistic pleiotropy plays a major role in structuring heritable variation. In a study of cranial morphology, Leamy et al (1999) found that most genes had modular effects with respect to the face and braincase. They found eight QTLs affecting only the face, eight QTLs affecting only the braincase, eight QTLs affecting the entire skull, and two QTLs with antagonistic pleiotropic effects between the face and braincase. Further, more anatomically detailed QTL studies are underway both on mouse and baboon crania.

In other studies of the developmentally complex mandible (Atchley & Hall 1991), we again found a modular pattern with most loci having their effects restricted either to the ascending mandibular ramus with its muscle attachment regions or to the tooth-bearing alveolus (Cheverud et al 1997, Ehrich et al 2003). These results strongly support the hypothesis of modular pleiotropy. Kenney-Hunt and colleagues (personal communication) are currently testing the modularity hypothesis across the whole skeleton in a hierarchical fashion, contrasting the appendicular and axial skeleton, and their component anatomical subsets.

Genetic variation in pleiotropy and differential epistasis

The question then arises as to how modular pleiotropy has evolved. If it has evolved through natural selection or genetic drift, the pleiotropic range of gene effects must itself be genetically variable. The pleiotropic effects of a gene can vary because of differential epistasis (Cheverud 1996b, Cheverud et al 2004). Epistasis occurs when the phenotypic effects of a gene vary depending on the genotypes present at another locus. If these epistatic effects differ from one trait to another, there is differential epistasis and genetic variation in the range and strength of pleiotropic effects (see Fig. 3).

The phenomenon of differential epistasis is well known in the genetics literature as the following examples illustrate. The abnormal abdomen mutation (aa) in *Drosophila* was long known in the laboratory and named for its effect on abdominal cuticle development, resulting in a juvenilized cuticle (Templeton et al 1985, 1993, Templeton & Johnston 1988, Hollocher et al 1992). In addition, 'aa' had pleiotropic

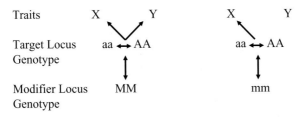

Differential Epistasis

FIG. 3.   Differential epistasis occurs when the pleiotropic range of traits affected by a locus varies depending on genotypes at a second, modifier locus. In this example the difference between homozygotes at the A locus affects both traits X and Y when the modifying locus has the MM genotype but the A locus effects are restricted to trait X when the modifying locus has the mm genotype. The mm genotype suppresses the effect of the A locus on trait Y.

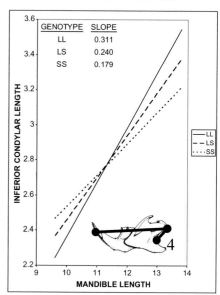

FIG. 4. Relationship quantitative trait loci (QTL) on chromosomes 17 and 11 affecting the relationship between mandible length and posterior corpus height and inferior condylar length, respectively. In each case, the slope of the regression of the specific mandibular region on total mandibular length changes depending on the genotype at the locus identified. These QTL are a source of genetic variation in pleiotropy. In each case the LL homozygote leads to a relatively small local mandibular region in short mandibles but leads to a relatively large region if the mandible is long.

effects on life history traits, including slow development time, higher early fecundity, and low adult survivorship. The same molecular mutation was discovered in a population in Hawaii living across a severe moisture gradient. However, its pleiotropic effects had been modified by epistatic interactions with other loci. The effect on the cuticle had been suppressed while the life history effects were maintained. The modified 'aa' mutation was selected for in dry conditions due to its adaptive life history effects. However, selection also acted to modify pleiotropy at the locus due to epistatic interactions by evolving a genetic background in which the deleterious juvenilized cuticle was suppressed because it is maladaptive in dry environments.

A second example comes from the agricultural research community (Geetha et al 1991, Moro et al 1996, Burnett & Larkins 1999). The opaque-2 maize mutant greatly enhances the amount of lysine in the endosperm. Lysine is an essential amino acid for humans and could enhance the nutritional value of corn. Unfortunately, the mutation also resulted in a thin seed coat that interfered with modern harvesting techniques. Researchers used backcrossing with recurrent selection to produce a thicker seed coat while maintaining the high lysine content effect of opaque-2. Through this selection, alleles at several loci that epistatically suppressed opaque-2's effect on the seed coat while maintaining its effect on lysine content increased in frequency. Again, the pleiotropic effects of the opaque-2 gene were modified by evolution at epistatically interacting loci.

Relationship quantitative trait loci

The phenomenon of differential epistasis can be seen in the effects of relationship QTLs (Fig. 4). Relationship QTLs are regions of the genome that affect the relationships between traits with alternate alleles producing stronger or weaker ties between phenotypes. Relationship QTLs provide genetic variation in the range of pleiotropy expressed at a locus (see Figs 2 and 3). We mapped relationship QTLs across the mouse genome affecting the relationship between individual parts of the mandible and overall mandibular size (mandible length; Cheverud et al 2004). We uncovered a total of 23 QTLs across 13 different chromosomes. About one-third of these QTLs also had direct effects on the mandible (Ehrich et al 2003). The principal of morphological integration was also expressed in the patterns of traits affected by pleiotropic relationship QTLs. At most genomic sites, the traits with a genetically variable relationship with mandible size consisted of sets of developmentally and functionally related traits. These loci produce genetic variation in the individuation of mandibular regions with respect to the mandible as a whole, or genetic variation in modularity. Such variations are critical for models of the evolution of developmental modules and the individuation of parts in evolution.

These results lead us to the hypothesis that the observed modular pleiotropic patterns are sculpted by natural selection so that functionally and developmentally related features are affected by module-specific genes, freeing the population phenotypic mean to evolve by masking deleterious pleiotropic effects.

Summary

I suggest that the most important relationship between development and evolution occurs through the effects of developmental processes and interactions on patterns of heritable variation as measured by the genetic variance/covariance matrix. We can predict patterns of heritable variation based on developmental relationships among traits, as realized through patterns of mutational variation and internal stabilizing selection. We found that developmentally related traits tend to be inherited together in a modular fashion because of modular pleiotropy and that genetic variation in the range of pleiotropic effects exists allowing for the evolution of module-specific gene effects.

Acknowledgements

This work was most recently supported by NSF grant BCS-0523305.

References

Ackermann RR, Cheverud JM 2004 Morphological integration in primate evolution. In: Pigliucci M, Preston K (eds) Phenotypic integration: studying the ecology and evolution of complex phenotypes. Oxford University Press, Oxford, p 302–319

Atchley WR, Hall BK 1991 A model for development and evolution of complex morphological structures. Biol Rev 66:101–157

Atchley WR, Xu S, Cowley DE 1997 Altering developmental trajectories in mice by restricted index selection. Genetics 146:629–640

Burnett RJ, Larkins BA 1999 Opaque-2 modifiers alter transcription of the 27 kDa g-zein genes in maize. Mol Gen Genet 261:908–916

Cheverud JM 1982 Phenotypic, genetic, and environmental morphological integration in the cranium. Evolution 36:499–516

Cheverud JM 1984 Quantitative genetics and developmental constraints on evolution by selection. J Theor Biol 110:155–172

Cheverud JM 1996a Quantitative genetic analysis of cranial morphology in the cotton-top (*Saguinus oedipus*) and saddle-back (*S. fuscicollis*) tamarins. J Evol Biol 9:5–42

Cheverud JM 1996b Developmental integration and the evolution of pleiotropy. Am Zool 36:44–50

Cheverud JM, Routman EJ, Irschick DK 1997 Pleiotropic effects of individual gene loci on mandibular morphology. Evolution 51:2004–2014

Cheverud JM, Ehrich TH, Vaughn TT, Koreishi SF, Linsey RB, Pletscher LS 2004 Pleiotropic effects on mandibular morphology II. Differential epistasis and genetic variation in morphological integration. J Exp Zool B Mol Dev Evol 302B: 424–435

Ehrich TH, Vaughn TT, Koreishi SF, Linsey RB, Pletscher LS, Cheverud JM 2003 Pleiotropic effects on mandibular morphology I. Developmental morphological integration and differential dominance. J Exp Zool B Mol Dev Evol 296B: 58–79

Geetha KB, Lending CR, Lopes MA, Wallace JC, Larkins BA 1991 Opaque–2 modifiers increase g-zein synthesis and alter its spatial distribution in maize endosperm. Plant Cell 3:1207–1219

Hollocher H, Templeton AR, DeSalle R, Johnston JS 1992 The molecular through ecological genetics of abnormal abdomen IV. Components of genetic variation in a natural population of Drosophila mercatorum. Genetics 130:355–366

Lande R 1979 Quantitative genetic analysis of multivariate evolution, applied to brain: body size allometry. Evolution 33:402–416

Lande R 1980 Genetic variation and phenotypic evolution during allopatric speciation. Am Nat 116:463–479

Leamy L, Routman EJ, Cheverud JM 1999 Quantitative trait loci for early and late developing skull characters in mice: A test of the genetic independence model of morphological integration. Am Nat 153: 201–214

Lynch M, Walsh B 1998 Genetics and analysis of quantitative traits. Sinauer Associates, Sunderland, Massachusetts

Marroig G, Cheverud JM 2001 A comparison of phenotypic variation and covariation patterns and the role of phylogeny, ecology and ontogeny during cranial evolution of New World Monkeys. Evolution 55:2576–2600

Moro GL, Habben JE, Hamaker BR, Larkins BA 1996 Characterization of the variability in lysine content for normal and opaque2 maize endosperm. Crop Sci 36:1651–1659

Olson EC, Miller RL 1958 Morphological Integration. University of Chicago Press, Chicago

Oxnard CE 1976 Primate quadrupedalism: Some subtle structural correlates. Yearb Phys Anthropol 20:538–553

Oxnard CE 1984 Order of man: A biomathematical anatomy of primates. Yale University Press, New Haven, Connecticut

Provine WB 1976 The origins of theoretical population genetics. University of Chicago Press, Chicago

Reidl R 1976 Order in living organisms. John Wiley & Sons, New York

Riska B 1986 Some models for development, growth, and morphometric correlation. Evolution 40:1303–1311

Rutledge JJ, Eisen EJ, Legates JE 1974 Correlated response in skeletal traits and replicate variation in selected lines of mice. Theor Appl Genet 45:26–31

Templeton AR, Crease TJ, Shah F 1985 The molecular through ecological genetics of abnormal abdomen in Drosophila mercatorum. I. Basic genetics. Genetics 111:805–818

Templeton AR, Johnston JS 1988 The measured genotype approach to ecological genetics. In: de Jong G (ed) Population genetics and evolution. Berlin:Springer-Verlag p 138–146

Templeton AR, Hollocher H, Johnston JS 1993 The molecular through ecological genetics of abnormal abdomen in Drosophila mercatorum. V. Female phenotypic expression on natural genetic backgrounds and in natural environments. Genetics 134:475–485

DISCUSSION

Hall: One of the complications of doing this on the mandible is that you had to do it on every cell population. Ideally, for the long bones, you would want to do it on each long bone of the limb. I assume you are going to start with just one, though?

Cheverud: We'll start with what looks promising from the pattern of effects. Some of those QTLs affected all of the long bones, and presumably all of their growth plates could be affected. We might sample some of their growth plates for that purpose. Others are specific to an individual bone or pair of bones. There is a nice femur/humerus QTL. This would restrict our examination of the fetus by the adult morphology we see. Sometimes the preparation is difficult: it can be hard to get both ends of the bone prepared properly. We use genetically replicable animals, and if the lower and upper growth plate are from different animals we will still be able to compare them.

Hall: That's nice to have that genetic homogeneity.

Wagner: If you hypothesize that the modular pattern of pleiotropy is evolving, then from what we understand from models that involve epistasis what is needed to evolve such genetic architecture is directional epistasis. Specifically, one would expect that between functional groups there is more negative epistasis than within the same group. Do you find any differences in the kinds of epistatic interactions among different groups of traits?

Cheverud: There is a lot of variety in the style of epistasis, and exactly which genotypes are involved in the interaction. This is a subject for future research. A lot of it seems to be that the epistasis will be present or absent for different traits. There will either be some or none. Some of this variability might have to do with suppressing epistasis on one of the traits, so it no longer shows the epistatic effect. Indeed this should result in the presence of more suppressing epistasis between functional groups.

Hallgrimsson: When rates change in the growth plate, do you expect the same developmental processes to be involved as might occur in a simple system, or can growth plates evolve different rates of growth by different developmental means? Can growth rate in the growth plate evolve by different developmental processes?

Cheverud: Yes, the growth plate can evolve by different developmental processes. Getting at this relies on making congenic lines where you isolate that one part of the genome from all of the rest. Then you could study in detail the basis for the growth rate variation at a particular QTL.

Hallgrimsson: You pointed out the increased size of the proliferative zone as one way of getting a faster rate of growth. When you do a cartilage-specific knockout of PTEN, which inhibits mitosis, you get the opposite: a faster growth rate, a shorter proliferative zone and a bigger hypertropic zone.

Cheverud: There are many different ways to get to the same place, again. It will be interesting to see whether some are used more than others.

Duboule: Over the past 20 years in developmental biology we have learned that in the vertebrae, which is a perfect example of a modular structure with several ossification centres, the mechanism that regulates the number of cells that condensate for a given ossification centre, and the mitotic index of these cells, all these mechanisms are either the same or tightly linked. Personally, I don't see how you can reconcile this with the fact that you can get to the same point with different mechanisms. I have the same question with regard to the tail. Getting a long tail by itself cannot provide any adaptive advantages. What may provide an advantage is what you do with your tail. You will be able to do different things if you have long vertebrae within the tail or a collection of short vertebrae. I don't think the question is whether you can get two long tails with different mechanisms; the question is whether these tails will achieve the same kind of performance. My answer would be no.

Cheverud: Relative to the selection criterion, tail length, the two populations did achieve the same performance. One of the major factors of tail length is giving off heat. It is hard to see how the size or number of vertebrae would affect that. If they were hanging from it or using it to generate force, I could see a tremendous possibility for difference. In relation to your first comment, if these different sources of developmental variability are indeed tightly correlated due to pleiotropy, our QTLs should affect several developmental processes rather than just one. This will be an interesting outcome from the gene mapping.

Stern: You are making strong inferences about the evolvability of these systems from experiments that are very unlike those that would occur in natural populations. You are starting with small populations and applying strong selection pressure on one aspect of the phenotype at a time. You are pulling out variants that may be able to construct superficially similar phenotypes in different ways. It is not at all clear that in natural populations selection is going to be trying to isolate one phenotype from the rest of the organism.

Hanken: A natural analogue of the phenomenon that Rutledge et al (1974) produced in the lab is seen in two genera of neotropical salamanders that comprise extremely elongated animals selected for living underground (Wake 1966). All have tiny reduced limbs and elongated trunks and tails. One genus (*Lineatriton*) has accomplished this by retaining the ancestral number of vertebrae (as seen in non-elongate genera), but each vertebra is elongated. The other genus (*Oedipina*) has done this by increasing the number of vertebrae. In this case the selection seems not to be for specialized tail function as much as for just an elongated body.

Cheverud: The difference in response could have functional consequences for future interactions with the environment. There would be a prospective effect of a developmental process that happens to have evolved.

Hanken: In Rutledge's experiment, selection for increased tail length yielded either more vertebrae or longer vertebrae; results could have gone either way. In the case of neotropical salamanders, there are many different genera yet most retain the same ancestral number of vertebrae. This appears to be a constrained feature, and *Oedipina*, the one genus that has increased the number of vertebrae, seems to have escaped the constraint. There is a biased distribution of the phenotype.

Cheverud: It would be interesting to see how this genus broke free of that constraint developmentally.

Brakefield: I have a thought experiment. In the mouse example, if you were to select using much larger sample sizes, would you expect to see a mixed strategy evolving? Posing this question a different way: Why don't they evolve longer tail length by doing it both ways together?

Cheverud: As you say, they could well evolve using both processes in a larger population with a smaller founder effect. Testing that though would be difficult because the real answer to that question is 67 cents a cage per day! Experiments are limited in size by practical considerations.

Wagner: Since these two traits are additive, the number of vertebrae and their size, as you start selecting and one of them is responding, the variance of this character is also increasing because of the scaling relationship between the variance and mean. This leads to symmetry breaking, so that the selection response of the character that responded first will continue to have a higher rate of response.

If you have multiple variables that add to the same phenotype and start selecting on one of them, it becomes a more effective target of selection. I suspect the fact that even in nature we don't see intermediates has to do with some kind of symmetry breaking.

Cheverud: This would occur with direct selection on the developmental processes themselves, not on the end phenotype.

Hall: That is interesting, because there is a distinct difference in timing between producing extra vertebrae (very early, when the somites are segmented) versus increasing length (much later in the life cycle).

Weiss: There are other precedents, such as Mackay's work on bristle number, Richard Lenski's work with bacteria and selection for malaria resistance in humans. They all show some shared mechanisms, but also different mechanisms. This supports the idea that evolution will pick whatever is there.

Oxnard: I want to return to the modularity issue. That work was done in rodents, which only have incisors and molars. Two neural crest cell populations were found relating to the two alveolar units containing those teeth. I have often wondered if two populations of cells were seen (i.e. two units) because intervening units were lost because of dental reduction. Our morphometric studies in primates (which have incisors, canines, premolars and molars) implied that there were four such alveolar units, one for each group of teeth. This seems to support the 'populations of cells idea', i.e. units or modules. However, even in living primates there are 'missing' teeth. There are many incisors missing (as many as six) if we are to believe the fossil record. Again, though living primates have two premolars; there are many more in some fossil primates. There are even many more molars in some fossil primates. If we had a primitive primate that had all of these, would we still see separate populations of crest cells and separate units of the alveolar process, or would we see something that really was just a spectrum? Would the situation it still be modular or would it be something more continuous, linear and quantitative? Is the modularity artificially produced by dropout of teeth? Could it have been linear originally?

Wagner: That is logically not consistent. If you consistently drop out say the third tooth, and then the eighth and sixth, then there has to be a modular system. If it was quantitative, you would just reduce them and still have a gradient of morphology.

Oxnard: Were there still modules in the bones when there was a full suite of teeth? There might well have been, particularly because the other parts of the mandible seem to have been modular. However, loss of teeth may impose an apparent modularity upon the bony processes which may actually have been continuous. Though our morphometric studies imply that each tooth is a unit, the bone in which they sit (and which is produced by totally different developmental processes) is more of a 'rubber' structure—that is, it can change shape in a continuous fashion.

Cheverud: It would have been modular if you could identify them as being these differentiated types of teeth.

Lieberman: How many strains of mice were the selection experiments done on?

Cheverud: For the tail experiment there were two up, two down.

Lieberman: It might have been interesting to do the experiments on different strains to see how the different backgrounds correlated.

Cheverud: You would certainly get different kinds of results as the QTL replication between different strain pairs is low.

Hall: It would be useful to look at the range and type of variation between the strains.

Bell: One of the things people are interested in is how labile pleiotropies are compared to individual traits. If pleiotropies are really stable features of lineages, this could constrain evolution over long periods. If they are labile, it doesn't really matter: they are not constraints.

Cheverud: The answer will be somewhere in the middle.

Lieberman: If you look at domesticated animals, there must be some conserved pleiotropies out there.

Cheverud: At least we found variation in them. If variation is there it can be selected on.

Budd: You touched on some general questions which have been much discussed in the Evo-Devo literature. These are topics like the relationship between modularity in the phenotype and genotype: which one comes first and how do they both evolve? You seem to be suggesting a way out of this problem with your hypothesis of how natural selection works to produce these pleiotropic effects and modularities. Do you see this as a general answer?

Cheverud: In general there is no answer to the question of whether the genotype or phenotype does it first: they both do it together at the same time, as part of the same system. The separation of the two into different parts, with one operating independently of the other, is wrong.

Weiss: If we look at modern human variation and the large number of studies that have been done on disease, they have shown that mapping results are hard to replicate. Usually a few chromosomal locations are replicated, but they end up explaining a small fraction of the variation. Another characteristic is that the more precisely defined the phenotype the fewer locations, but they are still mostly not replicable. The more sharply you define the trait the closer you get to something genetic. It makes a lot of sense: if you are getting close to something tied to a gene product, you are going to map to the same region again and again; if you are looking more broadly you aren't.

Cheverud: When formal studies have been done of the overlap of different strain crosses, replicated results haven't been obtained. Background effects seem to have a big role in which genes are displayed for natural selection or for phenotypic variation.

Lieberman: So often people interpret this as noise.

Oxnard: I was delighted to see the reference to our work on leaping. But the devil is in the detail. In primates, there are three different forms of leaping; different mechanical modes in different ecological situations associated with quite different anatomical adaptations. In each of these leaping types the species are from at least two widely separated phylogenetic groups. This immediately implies that there is an enormous amount of parallelism or convergence in the story.

References

Rutledge JJ, Eisen EJ, Legates JE 1974 Correlated response in skeletal traits and replicate variation in selected lines of mice. Theor Appl Genet 45:26–31

Wake DB 1966 Comparative osteology and evolution of the lungless salamanders, family Plethodontidae. Mem South Calif Acad Sci 4:1–111

Genetic networks as transmitting and amplifying devices for natural genetic tinkering

Adam S. Wilkins

BioEssays, 10/11 Tredgold Lane, Napier Street, Cambridge CB1 1HN, UK

Abstract. Genes never act in isolation but only through webs of functional connections called 'genetic networks'. The term 'genetic network', however, embraces a number of conceptually distinct entities. These include metabolic gene networks, protein 'interactomes', transcriptional networks, and the molecularly diverse networks that underlie development. That last category is the most complex and the one of most direct relevance to morphological evolution. It will be argued here that most microevolutionary 'tinkering' involves changes in such genetic networks. Unfortunately, the conceptual and technical problems in elucidating and characterizing these networks are substantial. In consequence, relatively few developmental genetic networks, and their evolutionary alterations, have yet been characterized in any detail. Nevertheless, the *generic* functional properties of these networks can help explain certain aspects of evolutionary change. In particular, the ways that development genetic networks act as both *transmitting* and *amplification devices* for genetic change will be described. The relationship of these properties to the sometimes puzzlingly rapid rates of organismal evolution will be discussed.

2007 Tinkering: the microevolution of development. Wiley, Chichester (Novartis Foundation Symposium 284) p 71–89

The relevance of genetic networks to the subject of this symposium, microevolutionary tinkering, is simple and direct, at least if one accepts a certain logical proposition. This syllogism runs as follows: If (A) all morphological/anatomical changes are the result of developmental processes and if (B) all developmental processes are underlain by genetic networks, then (C) evolutionary changes in morphology/anatomy must reflect changes of the underlying genetic networks, either in their components or structures or both.

Biologists are quite rightly suspicious of logical propositions as explanatory guides to the properties of living things. Two centuries of biological science, since the early 1800s, have revealed the endless capacity of living things to surprise us with their properties. Despite the general limits of applying logic to biological entities, however, the above syllogism is probably sound. Part 'A' is true as a matter of observation while part 'B', though less certain, is supported by everything that has been learned about genes and development in the past quarter century. If both 'A' and 'B' are valid statements, then 'C' must be true.

In effect, my central claim here is that *microevolutionary phenotypic changes are largely based on evolutionary tinkering with developmental genetic networks*. At first, this perspective seems quite different from the traditional view which emphasizes changes in the activities and functions of single, or small numbers of genes, to explain both developmental and microevolutionary changes. Indeed, the vast number of informative findings concerning the roles of specific genes in particular developmental processes might even seem to refute the need to think in terms of genetic networks. Yet, there is no fundamental conflict between this more traditional perspective and the network-centric view described above. In particular, mutations of genes that are part of the internal workings of networks—and most gene products are probably in this category (see below)—and which provide the basis for microevolutionary changes, create their effects through operational changes in the networks in which those genes reside. These changes may be subtle and quantitative or involve sharp changes in connectivity to 'downstream' genes but in either case, the resulting phenotypic effect must involve some ripple effect through the network. The only genes with microevolutionary potential in which one can safely ignore the network context are terminal cytodifferentiation genes in cells that have no further developmental participation, e.g. the keratins in skin. Yet a very large fraction of cytodifferentiation gene products influence further developmental events via intercellular interactions and these, too, therefore have to be regarded as internal components of networks.

The most dramatic effects, however, are associated with genes conventionally classed as 'regulatory'. When the crucial involvement of a key gene in the development of a trait is reported, whether it is that of a *Hox* gene or other transcription factor gene, a signal transduction gene, or some other gene product that directly affects the activities of many other gene products, the results *implicitly* highlight the workings of the genetic network through which that gene product exerts its effects.

An example is the loss of pelvic armature in certain freshwater species of stickleback derived from armoured marine forms (Shapiro et al 2004, Colosimo et al 2004). The intriguing conclusion of these studies is that this loss has involved apparently independent loss-of-function (l-o-f) mutations in the *cis*-controlling elements of a particular gene, *Pitx-1*, a transcription factor gene. In the absence of the expression of that gene activity in the relevant skeletal-forming tissues during development, the extensive pelvic armature of the ancestral forms does not develop. The gene mutation(s) responsible are believed to be in *cis*-regulatory enhancers for *Pitx-1*. Yet, while these observations explain the microevolutionary loss of pelvic armature, they highlight an even more intriguing evolutionary question: how did that skeletal armature evolve in sticklebacks in the first place? The transcription factor encoded by *Pitx-1*, after all, does not 'create' this skeletal structure on its own, but must work by activating the transcription of a whole suite of genes, many of which will undoubtedly prove to be other regulators. Just as the loss of *Pitx-1* activity must delete the functioning of a whole genetic network (or 'module', see below), the *origination* of the skeletal armature in the stickleback lineage must have involved the evolution of a whole new network/module, involving the creation or re-utilization (from other genetic contexts) of a whole suite of genetic interactions. It is relevant in this context that while much of the commentary on this work highlighted the role of *Pitx-1*, the initial work itself indicated the existence of at least four other loci in the microevolutionary loss of the pelvic armature (Shapiro et al 2004). This number is almost certainly a considerable underestimate of the actual number of genes required to build that skeletal superstructure in the marine stickleback species.

In thinking about genetic networks and development, a key point to remember is that *all* genes are parts of genetic networks, even the cytodifferentiation gene products whose activities or structural properties *directly* create the cellular phenotypes, namely the multitude of cytoplasmic metabolic enzymes, the numerous proteins of the cytoskeleton and the enzymes that manufacture the membrane lipids. These cytodifferentiation molecules are simply the ultimate outputs of genetic networks. No gene is expressed in isolation or autonomously of the regulatory machinery of the cell. Furthermore, in contrast to the early picture of gene regulation formed in the 1960s, based on the Jacob-Monod *lac* operon model, in which the great bulk of genes were deemed to be 'structural genes' and a tiny fraction 'regulatory', it is increasingly clear that, in eukaryotes, the greater fraction by far of all genes are involved in 'regulatory' processes (Wilkins 2007). These involve not just the control of transcription by proteins (and RNA molecules) but translational controls and RNA splicing. In addition, of course, there are the multifarious molecular mechanisms of signal transduction. Specific regulatory mechanisms may employ selective protein degradation or the post-translational modification of protein activities such as phosphorylation or acetylation, as well as those that chemically modify DNA directly, such as the methylation and demethylation of bases in DNA, their expression governed by the earlier-occurring events in the unfolding activities of each network.

Although this perspective on the centrality of gene networks has not yet permeated the evolutionary biology literature, the concept of genetic networks was implicit in and crucial to the theorizing of two early major figures, C. H. Waddington (Waddington 1940) and Sewall Wright (reviewed in Wright 1968). Their ideas about networks had little initial impact in evolutionary biology but the advent of molecular biology, and its findings about gene regulation, stimulated new thinking about genetic networks in the 1960s and 70s (Kauffman 1969, Britten &

Davidson 1969, 1971). The subsequent advances in developmental genetics in the 1980s helped move the network perspective into the working models of developmental biologists, e.g. Sander (1983).

Today, thinking about networks is widespread. It is, in fact, the focus of a whole new field, or at least set of approaches, termed 'systems biology'. This new focus on networks has been propelled in biology primarily by the development of a suite of diverse and ingenious technological developments in the past decade, but it has also been encouraged by the exciting discovery that many networks of various kinds, both within and outside of biology, have a similar ('scale-free') structure (reviewed in Barabasi & Oltvai 2004). Yet while the generic evolutionary features of scale-free networks have been well-described (Dorogovtsev & Mendes 2002, Barabasi & Oltvai 2004), comparatively little attention has been given to the actual relationships between the evolution of organismal phenotypes and that of the genetic networks that underlie those features, although a few books and papers have begun this exploration (Davidson 2001, Porter & Johnson 2002, Hinton et al 2003, Wilkins 2002, 2005). This comparative dearth of attention undoubtedly reflects the multiple difficulties of exploring and characterizing the genetic networks that underlie development. Since the collective set of problems involved in such work are not always clearly recognized, it may be of value to describe and enumerate them. The next section attempts a concise account.

'Genetic networks', a more elusive concept than at first appears

The difficulties inherent in dealing with the structure of genetic networks are of four general sorts. They are: (1) conceptual; (2) representational; (3) experimental; and (4) hybrid conceptual–empirical. While the experimental hurdles are widely appreciated, the others tend to be ignored. Let us begin with a conceptual matter, the question of definition.

(1) *The problem of defining the term.* Just what precisely do we mean by the term 'genetic network'? It is obvious that the character and significance of any network is intimately related to the kinds of elements of which it is composed and the nature of their interactions. Thus, the collective set of networks that have all been claimed to show a power-law or 'scale-free' organizational pattern of functional connections (e.g. the world-wide web, patterns of movie star acting partnerships, personal patterns of acquaintanceship in general, protein 'interactomes', metabolic systems) differ strikingly in their functional 'outputs', hence in their character and consequences, despite their generic organisational similarity.

What may be less apparent is that the term 'genetic network', which seems a precise, simple and unitary category, is itself an umbrella term for a large number of different entities. Thus, total protein interactomes for a species, namely the complete set of protein–protein interactions that have been mapped in such organisms as the fruit fly *Drosophila*, the nematode *Caenorhabditis elegans* and the budding yeast, *S. cerevisiae* differ from transcriptional networks, which portray the set of transcriptional interactions that take place within the organism. Both are categories of 'genetic network' and both are claimed to have the scale-free property but they differ not just in the obvious aspect of composition but in their functional properties and in what they predict. As a result, the protein interactome and the transcriptome network of a given organism are not isomorphous: neither can be converted into the other nor used to deduce the structure of the other—though each can supply hints about particular local aspects of network structure in the other kind and both can supply information of great value to those studying the evolution of morphological–anatomical features. Metabolic networks, in which each biochemical conversion step is carried out by a genetic product comprise a third class of genetic network.

One feature that diagrams of all three of these kinds of network (protein interactomes, transcriptional networks, metabolic networks) share is that they usually omit information about the cell/tissue locations and the developmental times at which the indicated interactions take place. In effect, these kinds of network diagram are *global depictions* of functional interactions. Each one may, in effect, be seen as a 'meta-network' for its particular organism (Wilkins 2005). Yet, to understand the developmental processes of any organism, one must know the particular molecular/genetic events that are occurring in given cells and cell groups at specific times. Hence, without supplemental information, the data provided by diagrams of these kinds of network cannot be directly used by either developmental biologists or evolutionary developmental biologists.

(2) *Representational inadequacies*. In contrast, information about space and time is intrinsic to and provided in the explicitly developmental genetic networks that Eric Davidson, Michael Levine and their colleagues, and others, have been analysing and showing (Davidson 2001, Davidson et al 2002, Jensen 2004, Levine & Davidson 2005, Stathopoulos & Levine 2005). These sorts of networks are our central concern here. In the diagrammatic depictions of these networks, the time line is indicated along one axis while spatial regions, e.g. particular groups of cells of a given type, can be delimited by coloured boxes or other means. If there are too many functional interactions to fit comfortably in one diagram, a series of diagrams depicting the events as a series of stages can be employed (Stathopoulos & Levine 2005). Whether single or multiple, however, the diagrams usually omit quantitative information about the interacting molecules—indeed, this information is often unknown—and the durations of the reactions, though the values of these parameters determine whether or not the interactions have a functional outcome.

A more fundamental problem is that the diagrams are static depictions of what are, intrinsically, highly dynamic processes. The diagrams thus, inevitably, give a somewhat distorted representation of the events. In principle, network diagrams could be animated. Yet, the information in the detailed static diagrams is difficult enough to take in and mentally process. Animating the sequences of change, particularly if the animation mimicked events in real time, would only compound the difficulties.

(3) *Experimental/practical problems.* The practical problems in elucidating genetic networks for development are equally serious. In particular, there is the problem of ascertainment bias: the depictions inevitably emphasize those interactions that have been looked for and discovered, on the basis of prior information. In particular, while these diagrams sometimes include information about signalling networks and interactions (Stathopoulos & Levine 2005), their focus has tended to be on transcriptional events, thus slighting other kinds of molecular interactions that may be just as important in triggering developmental switches. Indeed, the emphasis that has existed in evolutionary developmental biology for more than a decade on transcriptional enhancers as *the* primary sites of genetic change in developmental evolution (Carroll 1995, Akam 1998, Stern 2000, Davidson 2001, Carroll et al 2003) itself skews possible interpretations of how those evolutionary changes occur. There are good reasons for thinking that developmental evolution employs genetic alterations that affect every conceivable level of genetic regulation (Alonso & Wilkins 2005). This realization is leading to a new effort to construct databases of all the relevant molecular interactions in development (Zhong & Sternberg 2006).

In principle, these more inclusive approaches will supply the kind of information that is needed for evolutionary developmental biologists. The technical difficulties that remain are not trivial, however. In particular, assembling this kind of information is expensive in material resources, numbers of investigators and time. Furthermore, for the near-term future, at least, the kind of detailed elucidation of genetic networks in development that is needed will be easiest to obtain in the handful of model organisms that have been the backbone of molecular developmental analysis so far, namely, the fruit fly *D. melanogaster*, the nematode *C. elegans*, the sea urchin, *Strongylocentrotus purpuratus*, the frog *Xenopus laevis* and the mouse, *Mus musculus.* For analysis of evolutionary changes in the larger groups to which these species belong, comparable analysis will need to be extended to a whole range of non-model organisms, most of which are not yet susceptible to the kind of detailed functional/genetic characterization that is possible in some of these model organisms.

(4) *Combined conceptual-technical problems: delimiting modules and boundaries.* A more fundamental problem, which is both conceptual and technical, concerns the ways in which particular networks are parsed into their component functional units, so called 'modules', and in which the boundaries of both those modules and the networks themselves, both temporal and spatial, are determined. To assess the difficulty, one needs a rough working definition of 'genetic network' itself and I would

offer the following: *A genetic network consists of the unique set of genes and the unfolding pattern of interaction of their products in a developmental sequence to produce a specific morpho-logical–anatomical property, such as a cell, tissue, appendage or organ, or a colour or bristle pattern.*

Since a non-colonial multicellular organism develops continuously and in an integrated fashion from a fertilized egg cell, which itself is the product of a developmental sequence, it is difficult to set the precise upper temporal boundary of a network. (The developmental end-point is often much easier.) Furthermore, since each part of the developing embryo is co-ordinated in its development, by a series of signals, with other developing tissues, organs, etc., setting the spatial (cellular, tissue) boundaries of the network's domain is also not a trivial task. Nevertheless, it is clear that there is *some* degree of independence of different parts, particularly once an organ rudiment has been formed and, at the genetic level, these partially independent developmental entities are said to be governed by genetic 'modules'.

Modules are, in effect, pieces of genetic networks that have a quasi-autonomous behaviour with respect to each other, in terms of their outputs. Those outputs may, or may not, correspond to visually recognizable elements of the fully developed phenotype but are usually taken to be so. Thus, in principle, the development of a butterfly wing might be dissociable into modules for: (1) the overall size and shape of the wing; (2) the formation of individual scales; (3) the placement of elements of the colour pattern; and (4) the synthesis and deposition of particular pigments in particular areas. Each such module may be employed many times, e.g. the scale formation module, while others might operate once and for a short time only. Of course, they are *not* wholly independent entities, as indicated by the use of the prefix 'quasi' above, since they *must* be linked and co-ordinated in order to give the wild-type developmental sequence for the whole process. A crude schematic diagram of the relationship between the modules of a network is shown in Fig. 1. The key point in legitimating a particular part of a network as a module is how dissociable its output is from that of other modules. If two parts of the network, e.g. parts A and B in Fig. 1, can be independently evoked under different developmental conditions, then they can be legitimately described as separate modules.

Despite these various difficulties and ambiguities associated with the term 'genetic network', the concept is, as argued at the outset, essential for grappling with the nature of the underlying genetic changes in morphological evolution. Indeed, simply understanding some of the *formal properties* of network structure and operation can provide some clarity in understanding certain features of the evolutionary process that might, under the older gene-by-gene approach to evolutionary change, remain quite opaque. In another place, I have described how the network-perspective can help explain both the differences and similarities in 'homologous' yet clearly visibly different structures (Wilkins 2005).

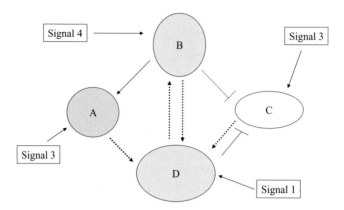

FIG. 1.   A hypothetical network composed of four 'modules', each of the latter consisting of a set of interacting genes (not diagrammed here). Every module is responsible for either a regulatory cassette of some kind (e.g. a signal transduction pathway), or a 'quantum' of phenotype (e.g. an increment of growth, a colour or colour pattern, bristles or nerve cells). Which particular modules are turned on and how they regulate each other will be a function of the mix of external signals, the effect of each module's output(s) on the other's, and the matter of timing of signals, when there are either competitive inhibitory interactions or positive synergistic ones between modules.

In the following two sections, I will describe how certain fundamental network properties can help explain the occasional rapid bursts of morphological evolution that have long been a contentious issue for conventional population genetic theory.

Pathways and networks as transmission devices

In principle, networks can be seen, and depicted, as arrangements of two kinds of 'structural motif', namely genetic pathways and the functional links between those pathways (Halfon & Michaelson 2002)[1], such links either being of the activating or inhibiting sort. These functional links between genes in a network constitute *transmission lines* of effect, once a gene activity in the upper reaches of the network ('upstream' steps) is triggered. Both networks and pathways should exhibit this property. To illustrate it, let us begin with pathways and the effect of genetic mutations in pathways, and then see how similar reasoning applies to networks.

---

[1] A third kind of structural 'motif' in networks is that of the feedback loop. These are either of positive or negative sign, respectively amplifying or stabilizing pathway activity, but for the purposes of this discussion, this class of motif can be ignored.

By definition, genetic pathways are linear sequences of gene action, involving at each step either an activation or inhibition of gene activity, with a fixed relationship between a specific triggering event (an 'input') and the final event (the 'output') of the operation of the pathway. The key point is that each pathway consists of an automatic, triggered sequence of events—in effect, each pathway is a transmission device. This will be the case whether the pathway consists solely of activation steps (type A), a mixture of activation and inhibition steps (type B), or solely inhibitory steps (type C) (see Fig. 1 in Wilkins 2005, and Fig. 2 below).

Figure 2 shows the effect of a severe loss-of-function (a 'null' or 'amorphic') mutation in an upstream gene in the three kinds of pathway. In all three cases, the effect of such a mutation is to *reverse the signs of activity of all downstream activities*. The crucial point is that epistatic interactions, which are typically envisaged as

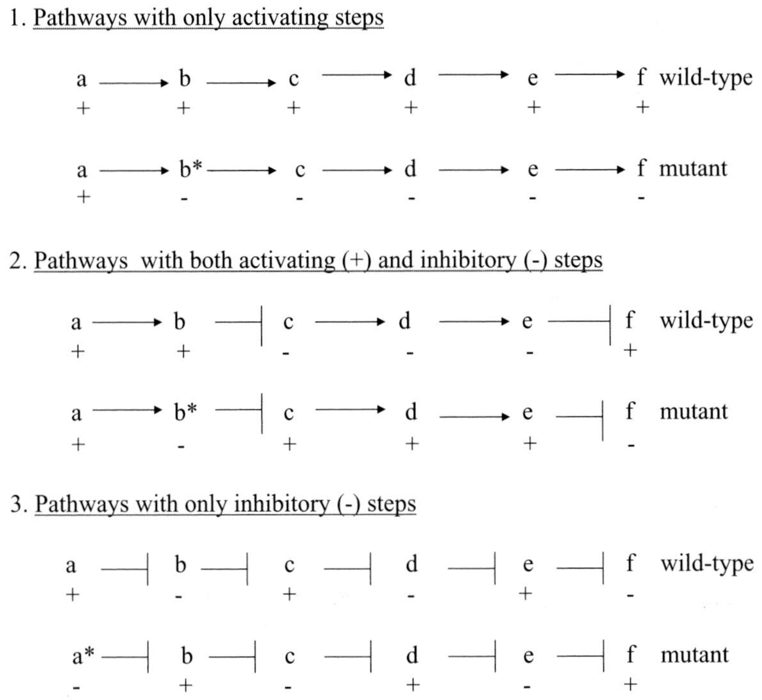

1. Pathways with only activating steps

```
a ——→ b ——→ c ——→ d ——→ e ——→ f  wild-type
+       +      +      +      +      +

a ——→ b* ——→ c ——→ d ——→ e ——→ f  mutant
+       -      -      -      -      -
```

2. Pathways with both activating (+) and inhibitory (-) steps

```
a ——→ b —⊣ c ——→ d ——→ e —⊣ f  wild-type
+       +      -      -      -      +

a ——→ b* —⊣ c ——→ d ——→ e —⊣ f  mutant
+       -      +      +      +      -
```

3. Pathways with only inhibitory (-) steps

```
a —⊣ b —⊣ c —⊣ d —⊣ e —⊣ f  wild-type
+      -      +      -      +      -

a* —⊣ b —⊣ c —⊣ d —⊣ e —⊣ f  mutant
-      +      -      +      -      +
```

FIG. 2.   The three kinds of possible pathway. Type A consists solely of positive (activation) steps; type B consists of a mix of positive and negative (inhibitory) steps; and type C consists solely of negative steps. Underneath each pathway is indicated the state of activation, either + or − , for each gene activity, when the pathway is activated, either for wild-type operation (top row) or when there has been a loss-of-function mutation, designated by an asterisk (*), in an upstream gene (bottom row).

involving closely functionally linked genes, can *extend down long sequences of gene action, namely genetic pathways*. An illustration of such genetically based reversal of sign of downstream gene activities is mimicked by a regulatory switch in the *C. elegans* sex determination pathway, a type C pathway, consisting solely of negative steps. In this pathway, change of sign of activity in the first step reverses all signs of activity downstream (Hodgkin 1980, Kuwabara & Kimble 1992). The key general point is that genetic pathways should act *as transmission lines for mutational effects*.

This reversal-of-sign effect for null mutants should also pertain to mutations that are only partial l-o-f mutations, namely hypomorphs, in those cases where critical gene activity (affected by the mutation) operates at a threshold close to that provided in wild-type. For gain-of-function (g-o-f) mutations, the outcomes should be more various and less predictable though some situations should also give this kind of sign reversal, in particular g-o-f mutations in genes that are normally inactive.

How does this property of pathways translate into the more complex situation of genetic networks? As noted above, networks can be visualized as sets of genetic pathways that are linked into reticulate structures. In this view, gene activities in one pathway that affect (either positively or negatively) a gene in another pathway in the network should transmit effects from upstream mutations in the first pathway to downstream activities in the second. Thus, the transmission effects described above should operate *across* such linked pathways although the precise effect will depend upon the nature of the functional link(s) and the nature of the mutation. For purposes of illustration, the simplest possible network, consisting of two pathways, with one functional link, will be described.

Imagine two hypothetical pathways, each consisting solely of activation steps, and each delivering one phenotypic output, when activated separately. Yet, if both are activated in the same cell(s), gene activity q in pathway 1 will inhibit gene activity x in pathway 2 (Fig. 3A). Hence, if both pathways are activated by their respective inputs (activating signals) only the output of pathway 1 will appear, that of pathway 2 will not. Thus, the outcome of the joint activation of the two pathways is not the sum of the two outputs but just the product of pathway 1. In effect, pathway 1 is epistatic to pathway 2. If one substitutes the term 'module' for 'pathway', then, in principle, one module can be epistatic to another—a relationship indicated in Fig. 1 by the bar symbol between modules D and C.

This conclusion, of course, is subject to all the necessary caveats about quantitative parameters that, in reality, would determine whether the concentration, cellular location, binding affinity, etc. of the inhibitory gene product in pathway 1 were sufficient to reduce the activity of the target gene product in pathway 2. In principle, however, the basic formalism should hold. Indeed, similar interactions have been observed in a study of metabolic modules in yeast, a property that has been termed 'modular epistasis' (Segré et al 2005).

A

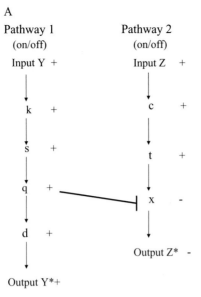

Pathway 1
(on/off)

Input Y +

k +

s +

q +

d +

Output Y*+

Pathway 2
(on/off)

Input Z +

c +

t +

x -

Output Z* -

B

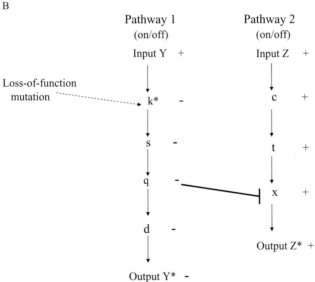

Pathway 1
(on/off)

Input Y +

Loss-of-function
mutation

k* -

s -

q -

d -

Output Y* -

Pathway 2
(on/off)

Input Z +

c +

t +

x +

Output Z* +

FIG. 3. (A) The wild-type situation for a simple network, one consisting of two pathways that are linked by a single functional (inhibitory) interaction. With inputs Y and Z, which activate pathways 1 and 2, respectively, the only output is Y*, that of pathway 1 because of inhibition of gene activity x (in pathway 2) by that of q (in pathway 1). This inhibition prevents pathway 2 from delivering its normal output, Z*. (B) The consequence of an upstream loss-of-function (l-o-f) mutation in the same network shown in (A). With such a mutation in gene k of pathway 1, the consequence is a reversal of all gene activity signs from k downstream. Now one does *not* get output Y* but only output Z*, the reverse situation to that of the wild-type. Depending upon the phenotypic 'quanta' delivered by the two respective pathways, the consequence of such a single l-o-f mutation in one pathway can be a visible, marked change in phenotype, even without any mutations in any of the gene products that directly create the visible cellular/tissue properties of that phenotype.

Now, imagine the situation in which there has been a loss-of-function mutation in gene k. This causes loss of activity of all genes downstream of gene k in pathway 1 and simultaneously lifts the inhibition of activity on gene q. The outcome of the effect of the mutation is to reverse the outcome of the wild-type situation. Now one obtains the output of pathway 2 but not of pathway 1. The situation is diagrammed in Fig. 3B. In effect, functional links between linear sequences of activity in networks, can make possible just the same kinds of transmission of effects (both in the wild-type organism and in mutants) seen above with pathways.

Networks as amplification devices

This line of thinking has a bearing on a major issue in evolutionary biology: the basis of those bursts of rapid morphological change that the fossil record bears witness to. Brought to general awareness of the scientific public by the model of 'punctuated equilibrium' (Eldredge & Gould 1972), such periods of accelerated evolutionary change have long been appreciated by palaeontologists (Simpson 1944). Such bursts of rapid change, however, could not readily be explained by classic population genetics, particularly of the schools of R.A. Fisher (Fisher 1930) and J.B.S. Haldane (Haldane 1932), which emphasized sequential genetic changes, each involving a phenotypic change of rather small effect. Indeed, Haldane was driven to explain the rapid mammalian radiation into the forms we know today, at the beginning of the Cenozoic, as due to an extraordinary burst of 'variation'[2]: 'Thus, the distinction between the principal mammalian orders seems to have arisen during an *orgy of variation* [emphasis added] in the early Eocene which followed the doom of the great reptiles . . .'

Although the actual nature of genetic changes in extinct forms cannot be determined, the evidence for rapid morphological evolution does not rely exclusively on the fossil record. It has also become apparent that speciation events, involving visible morphological change, can also take place in much shorter time spans, indeed, within the lifetime of human observers (Wiener 2005).

If one assumes that most complex morphological changes involve multiple gene effects, the transmission-of-effects inherent in genetic networks can contribute at least one element to the explanation of how such rapid morphological evolution occurs. If a given module of a network is responsible for some interpretable 'quantum' of phenotypic change, then, in principle, the *whole set* of genes in the module that are *upstream* of the final output can be the mutational target for altering that phenotypic quantum. If, for instance, there are 100 genes upstream of the

---

[2] Paleontologists have traditionally used the term 'variant' to denote the *visible* phenotypic form of the new type. Haldane, as a geneticist, however, was undoubtedly referring to the underlying genetic changes.

output—a not unreasonable figure given what is known so far of the better characterized genetic networks—then the mutational rate of the module will be 100 times that of the average gene in the module. In such circumstances, the mutation rate ceases to be such a tightly constraining feature on phenotypic/morphological evolution. Whether particular changes are promoted by standing genetic variation, which is usually held in check by some form of canalization (Gibson & Wagner 2000), or requires *de novo* mutation will, of course, affect the rate of change (compare the rates in the two boxes, below). To the extent that networks hold potential phenotype-changing standing variation, they act not just as transmission and amplification devices, but as *storage depots* for such variation, as well. As both the structure of networks and the nature of canalization become clearer, the precise ways in which networks act as storehouses of unexpressed variation should become clearer.

Two very rough calculations are shown for these two conditions, respectively, in Boxes 1 and 2. The most uncertain parameter in the calculation is the estimated percentage of base pair changes that might have some phenotype-changing potential. Here, I have used the figure 0.5%. Given the large amount of presumably non-functional DNA ('junk DNA'), this percentage might seem too high. New findings, however, about the ubiquity of mutations, both within and outside coding sequences, that can alter either splicing patterns (Garcia-Blanco et al 2004) or the

---

BOX 1   Standing genetic diversity in an average 'module'

The heterozygosity frequency at the nucleotide level has been estimated for different species in the range of 0.0001 to 0.01. Let's take the geometric mean of these values, or 0.001 (1/1000), as a standard figure. If one then estimates the average eukaryotic gene size as roughly 10 000 base pairs, then the frequency of heterozygous sites/gene will be

$$1/1000 \times 10000\,bp = 10 \text{ heterozygous sites/per average gene}$$

Let us say further that only 0.5% of these sites have a *potential effect* on that gene's activity, hence on its phenotypic consequences, when various buffering (canalizing) influences are released. This proportion would yield an average of 0.05 heterozygous sites/gene with some phenotype-altering potential (given appropriate environmental and genotype background.)

For a module of 100 genes, these figures lead to an estimate of five genes in the network *per individual* with *some potential* to alter the operation and output of the network. Even if the percentage of sites with some phenotype-altering potential is a 100-fold less, this figure would be 0.05 genes/network/individual, which is still a formidable genetic pool for a population of any reasonable size.

BOX 2   De *novo* mutations with network and phenotype-altering potentiality

The standard figure that is usually given for the base-pair mutation rate in eukaryotes is 10^{-6} mutations/gene/generation with a measured range, in different species, of 10^{-3}–10^{-7} mutations/gene/generation.

Take the standard mutation rate figure: then for a module consisting of 100 (10^2) genes, the frequency of new mutations in each module should be $10^{-6} \times 10^2 = 1 \times 10^{-4}$ or 1/10 000 new mutations per module per generation. If (see Box 1) 0.5% of all newly arising base-pair substitutions have some potential to alter module operation, then 1 in 500 000 individuals will have newly arising mutations with that potential. For animal or plant populations with large effect population size (Ne) values, even this seemingly low figure would permit adequate genetic resources for phenotype-altering change. Should the mutation rate be higher, say 10^{-4} mutations/gene/generation, then the figure will be 1 in 5000 individuals with new mutations of phenotype-altering potential in each 100 gene module.

---

activities of transcriptional enhancers (Wray et al 2002), are raising the estimated proportion of DNA that has potential mutability for phenotype-changing alteration. Even if the figure of 0.5% should prove 100-fold too high, the amount of hidden genetic variation within an average network with some phenotype-changing potential, in any population of non-miniscule size, would still be substantial.

Conclusions

Genetic networks are increasingly recognized as the genetic foundation of developmental processes. In the evolution of developmental processes, which is expressed as microevolution at the morphological level, the key genetic alterations involve both components of networks and, in the vast majority of cases, the evolution of new connectivities, the latter entailing either the gain of new functional connections or their loss. Since the in-depth characterization of developmental genetic networks for particular structures and processes has barely begun, with less than a dozen characterized to any completeness so far (see Stathopoulos & Levine 2005), the corresponding comparative studies of related networks, required by evolutionists, have lagged even further behind. Nevertheless, an understanding of the *generic* functional properties of networks can be useful for evolutionary biologists. That proposition has been illustrated here with a discussion of how network structure and operation helps explain the phenomenon of rapid evolutionary morphological change, a long-standing problem in evolutionary genetics.

References

Akam M 1998 Hox genes, homeosis and the evolution of segment identity: no need for hopeless monsters. Int J Dev Biol 42:445–461

Alonso C, Wilkins AS 2005 The molecular elements that underlie developmental evolution. Nat Rev Genet 6:709–715

Barabasi A-L, Oltvai Z 2004 Network biology: understanding the cell's functional organization. Nat Rev Genet 5:101–113

Britten RJ, Davidson EH 1969 Gene regulation for higher cells: a theory. Science 165:349–357

Britten RJ, Davidson, EH 1971 Repetitive and non-repetitive DNA sequences and a speculation on the origins of evolutionary novelty. Q Rev Biol 46:111–133

Carroll SB 1995 Homeotic genes and the evolution of arthropods and chordates. Nature 376:479–485

Colisomo PF, Peichel CL, Nereng K et al 2004 The genetic architecture of parallel armor plate plate reduction in threespine sticklebacks. PLoS Biol 2:635–641

Davidson EH 2001 Genomic Regulatory Systems. Academic Press, San Diego

Davidson EH, Rast JP, Oliveri P et al 2002 A genomic regulatory network for development. Science 295:1669–1678

Dorogovtsev SN, Mendes JFF 2002 Evolution of networks: from biological nets to the internet and WWW. Oxford University Press, Oxford

Eldredge N, Gould SJ 1972 Punctuated equilibrium: an alternative to phyletic gradualism. In: Schopf TJM (ed) Models in paleontology, Freeman Cooper, San Francisco p 82–115

Fisher RA 1930 The genetical theory of natural selection. Clarendon Press, Oxford

Gibson G, Wagner, GP 2000 Canalization in evolutionary genetics: a stabilizing theory? Bioessays 22:372–380

Garcia-Blanco MA, Baraniak AP, Lasda EL 2004 Alternative splicing in disease and therapy. Nat Biotechnol 22:535–546

Haldane JBS 1932 The causes of evolution. Longmans, Green and Co, London

Halfon MS, Michaelson AM 2002 Exploring genetic regulatory networks in development: methods and models. Physiol Genomics 10:131–143

Hinton VF, Nguyen AT, Cameron RA et al 2003 Developmental gene regulatory network architecture across 500 million years of echinoderm evolution. Proc Natl Acad Sci 100:13356–13361

Hodgkin JA 1980 More sex determination mutants of Caenorhabditis elegans. Genetics 96:649–664

Jensen J 2004 Gene regulatory factors in pancreatic development. Dev Dyn 229:176–200

Kauffman S 1969 Metabolic stability and epigenesis in randomly connected nets. J Theor Biol 22:437–467

Kuwabara PE, Kimble JE 1992 Molecular genetics of sex determination in *C. elegans*. Trends Genet 8:164–168

Levine M, Davidson EH 2005 Gene regulatory networks for development. Proc Natl Acad Sci USA 102:4936–4942

Porter AH, Johnson N 2002 Speciation despite gene flow when developmental pathways evolve. Evolution 56:2103–2111

Sander K 1983 The evolution of patterning mechanisms: gleanings from insect embryogenesis and spermatogenesis. In: Goodwin BC, Holder N, Wylie CC (eds) Development and evolution. Cambridge University Press, Cambridge

Segre D, Deluna A, Church GM, Kishony R 2005 Modular epistasis in yeast metabolism. Nat Genet 37:77–83

Shapiro M, Marks ME, Peichel CL et al 2004 Genetic and developmental basis of evolutionary pelvic reduction in threespine sticklebacks. Nature 428:717–723

Simpson GG 1944 Tempo and mode in evolution. Columbia, New York

Stathopoulos A, Levine M 2005 Genomic regulatory networks and animal development. Dev Cell 9:449–462

Stern DL 2000 Evolutionary developmental biology and the problem of variation: Evolution 54:1079–1091

Waddington 1940 Organizers and genes. Cambridge University Press, Cambridge

Wiener J 2005 Evolution in action. Nat Hist 114:47–51

Wilkins A 2002 The evolution of developmental pathways. Sinauer Associates, Sunderland, MA

Wilkins A 2005 Recasting developmental evolution in terms of genetic pathway and network evolution and the implications for comparative biology. Brain Res Bull 66:495–509

Wilkins A 2007 From the *lac* operon to 'gene recruitment': evolving ideas about gene regulation and its evolution. In: Hammerstein P Laublicher M (eds) Regulation. MIT Press, in press

Wray G, Hahn MW, Abouheif E et al 2002 The evolution of transcriptional regulation in eukaryotes. Mol Biol Evol 20:1377–1419

Wright S 1968 Evolution and the genetics of populations, vol 1. Genetic and biometric foundations. University Chicago Press, Chicago

Zhong W, Sternberg PW 2006 Genome-wide prediction of *C. elegans* genetic interactions. Science 311:1481–1484

DISCUSSION

Duboule: I am puzzled by the fact that even though from the beginning you made a distinction between pathways and networks, you actually exclusively used pathways to illustrate networks. If I understand that modifying pathways can modify the outcome of the pathway, one of the most important features of networks is that you can modify a pathway in the network without modifying the output of that particular pathway. You just need to add more connecting arrows between the different pathways. Networks are amplifying genetic modifications, but they can also buffer genetic modifications. There are as many examples of buffering as there are of amplification of genetic networks.

Wilkins: I disagree that I just talked about pathways. I was giving simple, two-pathway networks connected by a link. It is easiest to illustrate network properties with this kind of simple example. What I was saying certainly applies to networks and not just pathways. In addition, however, as you say, buffering is tremendously important. Probably, many of these networks in populations have a certain amount of genetic variance which under some circumstances can be released, but in the networks themselves there are all sorts of hidden buffering steps that would damp down the expression of this variance. One of my qualms about the way that Eric Davidson and Mike Levine investigate them, is that they are likely to miss out the buffering steps.

Jernvall: If we think about tinkering, it is more likely to involve existing network modification by eliminating a connection or altering the strength of interactions, rather than changing the topology of the whole network. In other words, if you are a 'gene network junkie' and see the world through the perspective of networks, would this be one operational way of making a distinction between what is tinkering and what is something more dramatic?

Wilkins: A lot of these regulatory connections are sensitive to the amount of gene products that then influences the activity of the next gene in the sequence. If you decrease the amount of a critical gene product below a certain level, you lose the connection. Sometimes, if you raise it, you then strengthen the connection such that this activator gene product does other things. I feel that many of the changes in networks involve quantitative changes which then involve some kind of subsequent linkage/connectivity changes. They wouldn't involve a total rewiring of the network but local connectivity changes with consequences that flow down.

Lieberman: This is what occurs through hidden genetic variation. You can have a network and some change in the environment and *voilà*.

Wilkins: My favourite example of a quantitative change is this effect in blind cave fish of raising their hedgehog secretion from the midline by a modest amount. With this modest increase a whole sequence of gene changes is set in train that leads ultimately to degeneration of the eyes.

R Raff: When I think about what happens with a hybrid embryo, my initial assumption is that this has to be explainable in terms of creating a new gene regulatory network. Then I fall into despair because I have no access to such a network. Then my heart leaps with joy because I realize that our experiments do actually give effects. This means that a lot of what we are seeing is pretty highly modularized, where we are perhaps not dealing with giant global networks all the time. It becomes possible to achieve some of the ends we are interested in without knowing anything about the global network.

Wilkins: I agree completely. Ultimately, however, if you really want to understand the evolution of every trait in your sea urchin, it would be helpful to know all the networks in detail. It would be useful to know which genes were being expressed and in what sequence, how this happens and how this leads to the cellular changes and so on. For much of the good experimental work that has been done, we don't need to know the networks in this kind of detail, but if we want to understand things in greater depth it would be good to have this information.

Budd: I am old fashioned and ill-informed enough to think about development in terms of hierarchies rather than networks. I wonder whether these network diagrams really assist us in our understanding, both of how development works and also how these systems actually evolved. They are very static.

Wilkins: Yes, they are static, and they also emphasize the regulator genes, whereas the business of developmental change involves changes in cells. But those aren't the genes that people tend to go after. They are less exciting, and are often gene products found in many different cells. But the way they are tweaked in their operation can make all the difference in terms of the cellular properties.

Budd: The idea is that development proceeds from the more general to the more specific. Do you get that impression from looking at regulatory networks? Or is it a property of the particular system in the sea urchin that gives rather a contrary view?

Duboule: I think a big problem with these networks is that some of them tend to start at the wrong extremity. Christiane Nüsslein-Volhard's beautiful work in the late 1970s was amongst the first to introduce arrows on slides whenever she and Eric Wieschaus found an epistatic relationship between genes and their products. This was the beginning of a powerful graphic representation, with arrows starting from the egg to the adult. However, we know today that there is probably as much variation and cooption in the early stages of embryogenesis as there is afterwards. This may be a serious problem, because if you start illustrating the network at either end of the phylotypic stage it can be misleading. The network must be started at the point where you assume it was an evolutionary start rather than a developmental start in order to make this network meaningful in an evolutionary context. It depends on the meaning you give to those arrows.

Lieberman: Those networks are often useless unless there is another network to compare it with. Scientists tend to focus on what they can observe. Sure, there is all this information, but it doesn't mean that it is relevant to the question being asked. To me, those networks are sometimes useful just to remind us of what is out there.

Bard: Perhaps the answer to the question of usefulness is the one Chou En Lai gave when he was asked about the significance of the French revolution: he replied, 'it is too soon to tell'. We are proceeding down this network path and trying it out; feeling our way. Networks may turn out to be wonderful and may help us understand the magic of development. But they may just disappear in a puzzle of complexity that we just can't untangle.

Wilkins: I would like to go back, for a moment, to the notion of hierarchy mentioned by Graham Budd and alluded to by Denis Duboule. The reason that the hierarchy notion got so embedded in people's thinking was partly because of the work of Christiane Nüsslein-Volhard, discussed by Denis, which seemed to indicate a nice clean hierarchy between different kinds of genes with discrete functions. It was well before we discovered how pleiotropic these genes are, and the fact that many parallel genes have multiple interactions. They are not just regulating or turning on the segment polarity genes, they are also doing things to the Hox genes. As soon as you have this kind of complexity, you lose a simple hierarchical relationship.

R Raff: If it were all network, then there is no gene that is over any other, yet when we do experiments we find we can measure effects. Thus, there are some places where linear events are occurring and there are hierarchies.

Stern: We know where those places are. I think the resolution of this whole problem is to remember that the genes are acting within cells. If you take these complex networks and look at what is happening in the single cell over time, most of the complexity gets stripped away. There are many fewer genes instructing a new subset of genes to do something. There is some autoregulation, and some genes are activated at later times in that same cell. If it is hard for us to interpret these networks, imagine how hard it would be for a cell! In fact, single cells are not interpreting these whole networks. They are interpreting only a small fraction of those networks at any point in time. It's when you lump them all together into one picture that it looks so impossibly complicated.

Duboule: One of the networks that Adam Wilkins showed is actually in a single cell. When you are talking about hierarchical relationships it is important to make a distinction between developmental hierarchies and evolutionary hierarchies.

Wilkins: What I was trying to say is that there are elements of hierarchy, but there may not be strict overall linear command-type hierarchies. That is an obsolete picture. I would also like to point out that the *Drosophila* segmentation and dorsoventral networks do not take place in a 'single cell'. Most of the genetic interactions take place after the syncytial phase of early embryogenesis.

Butterfly eyespot patterns and how evolutionary tinkering yields diversity

Paul M. Brakefield

Institute of Biology, Leiden University, P.O. Box 9516, 2300 RA Leiden, The Netherlands

Abstract. Eyespots are repeated elements in the wing pattern of butterflies. In the species-rich genus of *Bicyclus*, all eyespots are formed by the same developmental process. Artificial selection in *B. anynana* has explored how readily two of the eyespots can become different to each other. There is sufficient standing genetic and developmental variation in a single stock of this species for high flexibility in the responses for eyespot size; indeed selection over 25 generations in several directions of morphospace yielded phenotypes far beyond the variability found in the whole genus. In contrast, experiments on another eyespot trait, their colour composition, indicate that comparable flexibility occurs only along the axis of least resistance in which both eyespots change in the same direction. This result is reflected in both a clear difference in the developmental regulation of eyespot size and colour composition, and in the patterns of variability among species. Such research that integrates evolutionary genetics and Evo-Devo will eventually reveal how evolutionary tinkering occurs in both genetical and developmental terms, and will also explore the consequences of differences in evolvability for patterns of diversity.

2007 Tinkering: the microevolution of development. Wiley, Chichester (Novartis Foundation Symposium 284) p 90–109

The basic mechanisms of embryonic development in animals are extremely ancient and highly conserved. The induction and manipulation of major mutations in model organisms has identified many of the genes and their products that are central to development—to a surprising extent, the same 'toolkit' of transcription factor families, intercellular signalling pathways and so on, is used to build the highly divergent adult body forms of nematode worms, flies, fish, chickens and mice. It is becoming clear that much of the evolution of morphology has occurred not by the appearance of novel genes but rather by tinkering in the complex regulatory machinery of genes to change the position or timing of gene expression—old genes can learn new tricks in development (Carroll 2005).

The radiation in animal forms that has occurred in response to natural selection in different environments involves extensive tinkering in the mechanisms that build phenotypes. One goal of evolutionary developmental biology—Evo-Devo—is to explore the roles of development in evolution, and how these interact with the process of natural selection to shape biodiversity (Brakefield 2006). Differences in gene function during development generate the variations in phenotype that can then be sieved by natural selection to yield adaptive evolution. Evo-Devo as a field has been expanding to consider variation in natural populations, and is doing so in the context of morphologies that are relevant in an ecological context (see collection of papers in Brakefield & French 2006).

Our Evo-Devo work on butterfly wings has focused on how the potential for evolutionary change in complex traits is biased by the underlying genetic variation or developmental mechanisms. Although the associated concepts of evolutionary constraints arising from genetic channelling (Cheverud 1984, Lande & Arnold 1983, Schluter 1996, Blows & Hoffman 2005) or developmental bias (Maynard Smith et al 1985) have attracted much discussion, there have been few experimental tests of selection versus constraint (Brakefield 2006). Raup's (1967) classic example of how species of mollusc fill plots of potential morphological space for shell form as described by three parameters of growth, illustrates how maps of species occurrence in morphospace typically reveal that large parts of the potential space are not filled. However, such descriptions are not sufficient to confirm the involvement of genetic channelling or developmental bias along 'lines of least resistance' (Schluter 1996), a more experimental approach is needed.

Butterfly wing patterns

Evo-Devo is extending the work on the few model organisms of developmental biology to examine more subtle variation in morphology found within species and the differences characteristic of closely related species. This will reveal the extent to which genes with central developmental functions harbour segregating variation within natural populations that contributes to the developmental basis of morphological variation. This initiative is gaining momentum as more use is made in new systems of the understanding of development from the model organisms of developmental biology. The spectacular diversity of butterfly wing patterns is yielding a series of such studies. Whilst work on *Drosophila* has provided entries into developmental mechanisms in butterfly wings (e.g. Carroll et al 1994), species of this group are often more amenable to work in the field where they have fascinating biologies such as seasonal forms in *Bicyclus,* or mimicry in *Heliconius* (Beldade & Brakefield 2002).

Wing patterns in butterflies of the family Nymphalidae are made up of combinations of different pattern elements including series of colour bands, stripes and patches, and a series of marginal eyespots (Nijhout 1991); each series can be considered as a module. A reconstruction of a Nymphalid 'groundplan' shows these repeated elements arranged in columns along each surface of the fore- and hind-wings. Each wing is subdivided by wing veins into a series of wing cells, each of which has its own combination of these pattern elements. Whilst the development of the marginal eyespots is becoming understood, we still do not know how other pattern elements are formed, although the processes clearly involve different developmental genes (Beldade & Brakefield 2002).

The functional significance of butterfly eyespots

Eyespots resemble vertebrate eyes, being made up of concentric rings of scale cells around a central 'pupil', and with each ring having a different colour. Butterflies are subject to many predators that hunt by sight, especially lizards and birds (e.g. Brockie 1972, Chai 1996). Some eyespots function to intimidate predators, perhaps by taking advantage of looking like an eye (Blest 1957, reviewed in Stevens 2005). In these cases an eyespot is conspicuous and positioned towards the centre of a wing. They are usually hidden at rest and suddenly exposed on disturbance, frequently in combination with a ritualized behavioural display to startle any would-be predator (e.g. Vallin et al 2006). Other eyespots, including those in *Bicyclus,* are smaller, less conspicuous and closer to the wing margins. The latter can tear readily (DeVries 2002, Hill & Vaca 2004) so that if a predator attack is misdirected or deflected by the 'target' eyespots the butterfly can escape, albeit losing a small piece of wing tissue (Poulton 1890, Young 1979, Wourms & Wasserman 1985, Lyytinen et al 2004). In addition to their interactions with predators, eyespots can be important in sexual selection or mate choice (Breuker & Brakefield 2002, Robertson & Monteiro 2005).

Towards understanding evolutionary tinkering

Given a key innovation such as the ability to form a spot or series of spots on the wings of butterflies, how does the subsequent process of evolutionary tinkering occur in terms of genetic variation and changes in development? Transcription factor functions and intercellular signalling mechanisms central to embryonic development were apparently co-opted early in butterfly evolution to set up the capacity to make eyespots on scale-covered wings (Carroll et al 1994, Brunetti et al 2001, Reed & Serfas 2004). We can propose a scenario for these processes as shown in Fig. 1. This focuses on two traits of eyespots, their size and their colour composition. Our research will eventually show how close such a scenario is to reality, as well as reveal details of the underlying genes and developmental mechanisms. We have concentrated on the application of artificial selection in a laboratory model species, *Bicyclus anynana,* but are also now using a more comparative approach (Brakefield & Roskam 2006). A long term aim is to examine the extent

A

ELABORATION

ORIGIN?

TINKERING

and
radiation

B

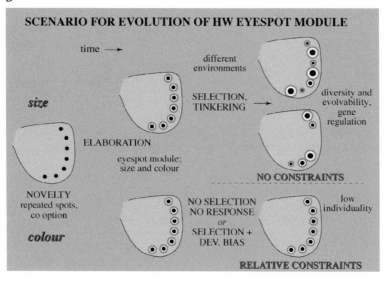

FIG. 1. Scenarios of eyespot evolution in nymphalid butterflies. (A) Specimens of a basal moth (left) and butterflies that illustrating a scenario of the evolution of the eyespot pattern on the ventral hindwing in the genera *Bicyclus* and *Mycalesis*. Evolution of the key innovation of making undifferentiated, small wing spots probably occurred in basal Lepidoptera. This was followed by a period of elaboration leading to a modular pattern with serial repeats of similar eyespots having both larger size and colour ring composition (with a representative species, *Mycalesis horsfeldi*, showing a pattern like this across both wings). Finally, species in different selective environments have diversified through a process of evolutionary tinkering. (B) Cartoon outlining the same scenario illustrating the evolution of the evolvability of eyespot size above, and of eyespot colour below. (Redrawn from Brakefield & Roskam 2006).

to which the mechanisms that generate variation in the eyespots influence the patterns of diversity found in the whole lineage (Brakefield 2006).

Artificial selection in certain organisms can be a powerful tool for uncovering potential sources of bias in evolutionary change. However, if artificial selection is to help inform us about the adaptive evolution of complex morphologies, the sources of variation targeted in such experiments must have parallels with the process of generating phenotypic variation and the responses to natural selection in different environments in the wild. Revealing the extent to which phenotypes yielded by artificial selection in a laboratory model resemble those of related species in terms of the underlying genetic and developmental changes is an exciting challenge for the future. Artificial selection can be highly targeted at specific traits, and is likely to be less influenced by pleiotropic effects on fitness than are responses to natural selection. So the success of our approach is likely to go back to the point made above, about whether variation segregating within natural populations provides the bases for the differences observed among species. Species frequently exhibit substantial spatial and temporal variability in different local environments, and they may have leaky genetic bounds. Whilst perhaps some or even many of the alleles that differentiate related species will prove to be 'private', the associated genes will have much more in common. It will also be fascinating to determine the complexity of differences in morphological traits across species both in terms of the numbers of underlying alleles and the size distribution of their individual effects. The answers to such issues are likely to depend on the traits and range of environments concerned and, in particular, on the extent to which such traits lack strong pleiotropic interactions with other traits or modules.

Novelty/elaboration/tinkering

Primitive Lepidoptera (as well as more derived species) frequently show a series of small, undifferentiated spots, towards their wing margins (see Fig. 1). Each spot is positioned between a pair of the major proximal-distal wing veins that divide a wing up into wing cells. This patterning probably reflects a 'simple' phenotypic module comparable with the early novelty that saw the origin of the developmental capacity to make patterns of repeated spots in each wing cell (Fig. 1). Such a simple pattern became much elaborated in the Nymphalidae to yield a series of repeated marginal eyespots, each with concentric rings of colour. The eyespots now range widely in size and colour—evolutionary tinkering has yielded eyespot complexity and diversity (Fig. 1) (Brunetti et al 2001, Beldade et al 2003, 2005). This process of elaboration and tinkering also results in changes in evolvability. How has this been achieved, and what are the consequences for the occupation of eyespot morphological space?

The eyespots of *B. anynana* are all formed by the same developmental pathway (Brunetti et al 2001, Beldade & Brakefield 2002, Reed & Serfas 2004). Transplantation experiments performed in early pupae show they are formed around groups of organizing cells called foci; transplanting an eyespot focus to a novel site results in formation of an ectopic eyespot around the graft (French & Brakefield 1995). Establishment of the foci occurs in late larvae, and then in early pupae each focus sets up a gradient in the surrounding epithelial cells, presumably via one or more signalling, diffusible morphogens. These cells then respond to the gradient and depending in some way on concentration, become fated to synthesize a particular colour pigment just before adult eclosion.

The seven eyespots on the ventral surface of the hindwing of *B. anynana* all express the same developmental genes, including *Distal-less*, *hedgehog*, *engrailed*, *spalt* and *Notch*, and at the same stages in eyespot formation (Brunetti et al 2001, Beldade et al 2002, 2005, Reed & Serfas 2004). Typically, mutant alleles established in laboratory stocks also affect all eyespots. Moreover, artificial selection experiments targeted on the size or colour of a single eyespot produce highly correlated responses in other eyespots, especially those on the same wing surface. The shared morphogenesis led us to design experiments to examine the potential flexibility of the repeated eyespot elements to evolve in different directions in trait space.

Results of artificial selection experiments on eyespot size

The dorsal forewing of the wild-type *B. anynana* shows a small anterior eyespot and a large posterior eyespot, a pattern roughly in the middle of trait space (Fig. 2). Rapid responses to 'upward' and 'downward' artificial selection on the size of the posterior eyespot occurred in 'high' and 'low' lines (Monteiro et al 1994). These were accompanied by strongly correlated responses in the anterior eyespot. In contrast, eyespot colour-composition changed very little, indicating low genetic correlation between the two traits (see also Monteiro et al 1997). Transplantation experiments with the eyespot focus traced much of the difference in eyespot size between the lines to differences in the activity of the signalling cells. Thus foci donated by pupae of the high line consistently yielded larger ectopic eyespots whether grafted into pupal hosts of the large- or the small-spotted butterflies (Monteiro et al 1994).

A later experiment explored whether a phenotype in which one eyespot was smaller and the other larger could be produced as readily by artificial selection. Replicated lines were established from the same founder population and selected towards each of the four corners of trait space (see Fig. 2); that is, along both directions of the 'coupled' axis following the shared genetics and development, and along both directions of the 'uncoupled' axis orthogonal to this proposed genetic

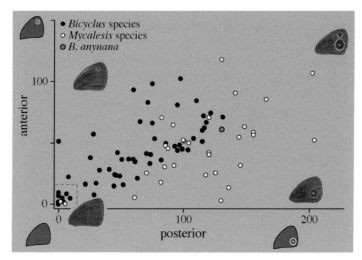

FIG. 2.   The occupancy of morphological space for the relative size of the two dorsal forewing eyespots of the butterfly *Bicyclus anynana* compared to the variation among species of this genus and of *Mycalesis*. The four images of the wing to each corner of the morphospace are representative examples of the wing pattern after 25 generations of artificial selection in *B. anynana* in these directions starting from the wild-type for this species as depicted by the semi-closed circle in the middle of the diagram. The four wings are placed in roughly the correct position in the depicted trait space. Circles show the positions of the mean patterns of the size of the same eyespots for different species of the African genus *Bicyclus* (closed symbols) and of the closely related Asian genus *Mycalesis* (open symbols). The dotted square encloses species for which both eyespots are very small or absent, and frequently difficult to measure. Reproduced from Brakefield & Roskam (2006) Am Nat 168 (Suppl 6):S4–S13.

line of least resistance and plane of developmental bias. Selection occurred over 25 generations (Beldade et al 2002).

As expected, selection either 'up' or 'down' the 'coupled' axis of shared development produced rapid responses, with butterflies eventually either having no eyespots or two very large ones. These morphologies are completely different to any present in the base population. However, populations along the other 'uncoupled' axis, orthogonal to the axis of shared development, also responded rapidly to selection, eventually producing phenotypes in which one eyespot was very large and the other absent or very small (see superimposed wings towards the corners of Fig. 2). We concluded that this pattern in relative size of the two eyespots behaved in a highly flexible manner with a high evolvability in all directions (Beldade et al 2002). We also suggested that this capacity for independent evolution was the product of a long legacy of natural selection and evolutionary tinkering leading to morphological diversity among species and to the appropriate genetic and developmental properties of the standing variation for eyespot formation at different sites in the wings (Beldade et al 2003).

The flexibility observed in the responses to selection on the pattern of eyespot size does not necessarily imply zero potential effect of developmental bias or genetic chanelling on eyespot diversity (Brakefield & Roskam 2006). An initial examination of the genus *Bicyclus* suggested that although species occurred with a large anterior eyespot and no posterior eyespot, there were none showing the reversed pattern (Beldade et al 2002). A more quantitative analysis has now been made of the occupancy of this morphological space using measurements of specimens from museum collections (Brakefield & Roskam 2006). This analysis used 69 species of *Bicyclus* from sub-Saharan Africa, together with 40 species of the closely related genus, *Mycalesis*, that extends throughout Asia into Australia. The pattern of occupation of morphological space within the lineage can then be compared to the results of the artificial selection based on the standing genetic variation at a single locality for a single species (Fig. 2).

Many species of *Bicyclus*, including *B. anynana*, clearly lie along the coupled axis that appears to reflect a 'line of least resistance' in Schluter's (1996) terminology. There are, however, also species that tend more towards having a large anterior eyespot combined with a small posterior eyespot. The rest of the morphological space, not only that representing the opposite pattern in relative size but also in the direction of both eyespots being very large, is not occupied by any (extant) species of *Bicyclus* in the survey. Since artificially selected lines from a single laboratory population of one species readily traversed this morphological space, we must conclude that such patterns are absent in nature because of the way in which natural selection has operated in the past rather than because of any strong developmental constraints.

This interpretation is given more weight by extending the comparison to the closely related genus of *Mycalesis* (Fig. 2). Here there is a (weaker) trend towards following a parallel genetic line of least resistance, but many species are displaced towards the pattern with large posterior eyespot and small anterior eyespot (lower right) rather than the opposite pattern as in *Bicyclus* (top left). Some recent experimental work suggests that these dorsal anterior eyespots are involved in sexual selection and mate choice (Breuker & Brakefield 2002, Robertson & Monteiro 2005, Brakefield & Roskam 2006). It is then tempting to suggest that the mechanisms of courtship and mate choice in *Mycalesis* may differ, leading to more emphasis on the posterior part of the dorsal forewing (see Brakefield & Roskam 2006) and another pattern of occupancy of morphospace. Whatever the explanation, details of the process of evolutionary tinkering and of eyespot formation are unlikely to account for the overall pattern of occupancy of morphspace for eyespot size.

Evolutionary tinkering and other eyespot traits

We have also examined the developmental genetics of another trait of the posterior forewing eyespot, namely colour composition. Artificial selection rapidly yielded either 'gold' or 'black' butterflies with broad or narrow outer gold rings, respectively (Monteiro et al 1997). As for eyespot size, the anterior dorsal eyespot showed a highly co-ordinated response in colour, whilst the two traits behaved independently (see also Beldade et al 2002, 2003). Although these observations on genetic variances predict comparable evolutionary responses to a given mode of natural selection, caution is needed since the two traits differ rather cleanly in development. Thus whilst the earlier transplants showed that large eyespots result primarily from 'strong' focal signals, comparable experiments using the 'gold' and 'black' lines traced no effect to the focus, but only to propagation of the signal gradient or the threshold responses of the surrounding epithelial cells to the focal signals that eventually form the colour rings (see Beldade & Brakefield 2002). Thus, there is support for two divergent developmental aspects of these traits: (1) eyespot colour is a wing-level property (gradient propagation and threshold responses in the wing epithelial cell layer), whilst size is dependent largely on the localized focal signal in the wing cells between each pair of wing veins; and (2) because size is also influenced partly by response properties, there may be more developmental 'options' for changing size than colour. We can now begin to examine the evolutionary consequences of such differences as well as their origin.

There is variability among species of *Bicyclus* and *Mycalesis* in the breadth of the outer gold colour ring of their marginal eyespots. However, although there are species characterized by narrow, intermediate or broad gold rings, we have seen no species with a wide range of colour composition amongst the different eyespots on a single wing surface (see Brakefield & Roskam 2006). Thus this variability differs sharply from that for eyespot size, where the patterns in relative size frequently differ dramatically across species. These observations suggest that it may be more difficult to uncouple the colour of two eyespots on the same wing (i.e. one 'gold' and the other 'black') by means of artificial selection on the standing genetic variation of a single species. Such experiments have yielded dramatically different results to those with eyespot size (Allen CE, Beldade P, Zwaan BJ & Brakefield PM, submitted manuscript). In particular, whilst comparable responses to those for size are observed when both eyespots are selected in the same direction, antagonistic selection produces no novel phenotypes of gold-black or black-gold.

Perspectives

Developmental studies on *Bicyclus* eyespots have suggested different mechanisms of pattern determination for eyespot size and colour (Monteiro et al 1994, 1997).

Artificial selection experiments are revealing differences in the evolvability of the pattern of two or more eyespots for these different traits that can be traced to the properties of morphogenesis. If the early results from a combination of the artificial selection and comparative analyses across extant species are confirmed, a relative lack of genetic variation to facilitate an 'uncoupling response' for eyespot colour in contrast to size could be accounted for by: (1) no history of selection in different environments for butterflies with a combination of some eyespots with 'narrow', and others with 'broad' rings; (2) a more absolute developmental constraint for eyespot colour perhaps relating to the wing-level property of responses to focal signals; or (3) a combination of these two effects working in harness to make it less likely for any substantial evolution of evolvability in colour for different eyespots.

Figure 1B illustrates these scenarios. We argue that early in the radiation of the Lepidoptera, an evolutionary novelty that involved co-option or recruitment of the hedgehog signalling pathway late in development yielded the pattern of serial repeats of small, undifferentiated spots (see Brunetti et al 2001). Such patterns are found in many Lepidoptera, including basal groups. There followed a period of evolutionary elaboration involving further recruitment of gene function to yield the module of repeated structures, each with a potentially larger and more structured colour pattern element. This pattern is illustrated in the central column of Fig. 1A by *M. horsfeldi*. Finally, species of nymphalid butterflies in different environments of predator pressure, of light and resting backgrounds, and of communities of conspecifics, have evolved through extensive tinkering in the developmental pathways the ability for independent evolution in the serial eyespot elements, at least for eyespot size. For some reason, perhaps accounted for by a combination of selection history and developmental bias, this process in *Bicyclus* and *Mycalesis* is substantially less advanced for eyespot colour than size. Future integrative work will examine the basis for this difference in terms of both genetic variances and developmental mechanisms (Beldade et al 2005).

Casting the net wider in comparative terms will reveal more about evolutionary tinkering. For example, we have established a species in the laboratory from another closely related genus, *Heteropsis*, that has radiated on the island of Madagascar (Torres et al 2001). In some species of this lineage, the large posterior forewing eyespot (but not the anterior) has evolved an extremely broad gold ring and also become more centrally positioned on the wing. Early work on *H. iboina* has suggested that the change in colour is due to the combination of an eyespot pattern element with a patch element of gold colour (S. Saenko et al, unpublished data). These changes are also associated with recruitment of a novel 'startling' function for the new posterior eyespot that incorporates both changes in wing movements and in behaviour. A similar function and display is observed for eyespot-like markings on the hindwings of species of *Parnassius* in the family Papilionidae

(Descimon et al 2002). In at least one species, *P. apollo*, these eyespots appear to be formed solely through modifications of colour patches (S. Saenko, H. Descimon et al, unpublished data). These as yet sketchy examples illustrate how fascinating it will be to unravel the complete pathways of evolutionary tinkering that have resulted in the spectacular diversity in butterfly eyespot morphology and ecology we see today.

Acknowledgements

Many researchers have worked on butterfly eyespots over the years, and this paper would not have been possible without their contributions. I also thank Gunter Wagner for his perceptive comments on the manuscript.

References

Beldade P, Brakefield PM 2002 The genetics and evo-devo of butterfly wing patterns. Nat Rev Genet 3:442–452

Beldade, P, Koops K, Brakefield PM 2002 Developmental constraints versus flexibility in morphological evolution. Nature 416:844–847

Beldade, P, Koops K, Brakefield PM 2003 Modularity, individuality, and evo-devo in butterfly wings. Proc Natl Acad Sci USA 99:14262–14267

Beldade P, Brakefield PM, Long, AD 2005 Generating phenotypic variation: prospects from 'evo-devo' research on *Bicyclus anynana* wing patterns. Evol Dev 7:101–107

Blest AD 1957 The function of eyespot patterns in the Lepidoptera. Behaviour 11:209–255

Blows MW, Hoffmann AA 2005 A reassessment of genetic limits to evolutionary change. Ecology 86:1371–1384

Brakefield PM 2006 Evo-devo and constraints on selection. Trends Ecol Evol 21:362–368

Brakefield PM, French V 2006 Evo-devo focus issue (editorial to special issue). Heredity 97:137–138

Brakefield PM, Roskam JC 2006 Exploring evolutionary constraints is a task for an integrative evolutionary biology. Am Nat 168 (Suppl 6):S4–S13

Brockie RE 1972 Evolutionary studies on *Maniola jurtina* in Sicily. Heredity 28:337–346

Breuker CJ, Brakefield PM 2002 Female choice depends on size but not symmetry of dorsal eyespots in the butterfly *Bicyclus anynana*. Proc R Soc Lond B Biol Sci 269:1233–1239

Brunetti CR, Selegue JE, Monteiro A, French V, Brakefield PM, Carroll SB 2001 The generation and diversification of butterfly eyespot color patterns. Curr Biol 11:1578–1585

Carroll SB 2005 Evolution at two levels: on genes and form. PloS Biol 7:1159–1166

Carroll SB, Gates J, Keys DN et al 1994 Pattern formation and eyespot determination in butterfly wings. Science 265:109–114

Chai P 1996 Butterfly visual characteristics and ontogeny of responses to butterflies by a specialized tropical bird. Biol J Linn Soc Lond 59:37–67

Cheverud JM 1984 Quantitative genetics and developmental constraints on evolution by selection. J Theor Biol 110:155–171

Descimon H, Elmquist H, Pierrat V 2002 Le determinisme genetique du graphisme alaire chez *Parnassius apollo* L. (*Lepidoptera: Papilionidae*): le cas de l'aberration *wiskotti* Oberthur. Linneana Belg 18:243–254

DeVries PJ 2002 Differential wing toughness in distasteful and palatable butterflies: direct evidence supports unpalatable theory. Biotropica 34:176–181

French V, Brakefield PM 1995 Eyespot development on butterfly wings: the focal signal. Dev Biol 168:112–123

Hill RI, Vaca JF 2004 Differential wing strength of Pierella butterflies (Nymphalidae, Satyrinae) supports the deflection hypothesis. Biotropica 36:362–370

Lande R, Arnold SJ 1983 The measurement of selection on correlated characters. Evolution Int J Org Evolution 37:1210–1226

Lyytinen A, Brakefield PM, Lindström L, Mappes J 2004 Does predation maintain eyespot plasticity in *Bicyclus anynana*? Proc R Soc Lond B Biol Sci 271:279–283

Maynard Smith J, Burian R, Kauffman S et al 1985 Developmental constraints and evolution. Q Rev Biol 60:265–287

Monteiro A, Brakefield PM, French V 1994 The evolutionary genetics and developmental basis of wing pattern variation in the butterfly *Bicyclus anynana*. Evolution 48:1147–1157

Monteiro A, Brakefield PM, French V 1997 Butterfly eyespots: the genetics and development of the color rings. Evolution 51:1207–1216

Nijhout HF 1991 The Development and Evolution of Butterfly Wing Patterns. Smithsonian Institute Press, Washington

Poulton EB 1890 Colors and markings which direct the attention of an enemy to some non-vital part, but which are not attended by unpleasant qualities In: Poulton EB (ed) The colours of animals. 2nd edn. London: Kegan Paul, Trench, Trubner & Co, p 204–209

Raup DM 1967 Geometric analysis of shell coiling: Coiling in ammonoids. J Paleontol 41:43–65

Reed RD, Serfas MS 2004 Butterfly wing pattern evolution is associated with changes in a Notch/Distal-less temporal pattern formation process. Curr Biol 14:1159–1166

Robertson KA, Monteiro A 2005 Female *Bicyclus anynana* butterflies choose males on the basis of their dorsal UV-reflective eyespot pupils. Proc R Soc Lond B Biol Sci 272: 1541–1546

Schluter D 1996 Adaptive radiation along genetic lines of least resistance. Evolution 50: 1766–1774

Stevens M 2005 The role of eyespots as anti-predator mechanisms, principally demonstrated in the Lepidoptera. Biol Rev 80:573–588

Torres E, Lees DC, Vane-Wright RI, Kremen C, Leonard JA, Wayne RK 2001 Examining monophyly in a large radiation of Madagascan butterflies (Lepidoptera: Satyrinae: Mycalesina) based on mitochondrial DNA data. Mol Phylogenet Evol 2001 460–473

Vallin A, Jakobsson S, Lind J, Wiklund C 2006 Crypsis versus intimidation—anti-predation defence in three closely related butterflies. Beh Ecol Soc 59:455–459

Wourms MK, Wasserman FE 1985 Butterfly wing markings are more advantageous during handling than during the initial strike of an avian predator. Evolution 39:845–851

Young AM 1979 The evolution of eyespots in tropical butterflies in response to feeding on rotting fruit: an hypothesis. J NY Ent Soc 87:66–77

DISCUSSION

Stern: In some of the ablations with the new species of *Heteropsis* from Madagascar that has a forewing eyespot with a very broad outer yellow ring, you seem to get rid of most of the yellow, in others you don't. Is it possible that most of that yellow is dependent on the eye spots? That is, the signal is required earlier than for the black ring.

Brakefield: It is possible. My comments were based on the first 10 or 20 ablation experiments we have done (unpublished). Perhaps if we can really establish the new species in the laboratory then in a few years we will have solved it. I think there is something interesting going on here in terms of combining these two pattern elements, namely one based on an eyespot, and the other on a colour patch giving a black eyespot essentially imbedded in a more or less round yellow or gold colour patch. However, we have not yet excluded the possibility that the complete pattern reflects solely a modification of an eyespot.

Lieberman: Do you get much asymmetry?

Brakefield: No.

Lieberman: Do the selection experiments change this?

Brakefield: If they do, it is a subtle effect. We have targeted left-right asymmetry by artificial selection. We had a graduate student who spent four years doing selection experiments on fluctuating asymmetry. We thought we had a great system because there is a *Spotty* mutant that adds in two novel eye spots to the pattern on the forewing. These novel eye spots have much higher asymmetry than the flanking wild-type eyespots. We set up selection experiments to target the difference in asymmetry between the mutant eye spots and the flanking wild-type eye spots. Nothing happened.

Jernvall: Going back to the variation, if you look in non-selected populations do you see variation along that axis?

Brakefield: There is an ellipse of phenotypic variation along that axis. This needs to be taken into account in looking at rates of response along the different axes. We have done this properly. For both colour composition and for size, the phenotypic variation is also larger along that axis. This could also account for, or contribute to, the bias in natural populations. Thus, for a given amount of natural selection, the time window required to move from wild-type pattern a certain distance along different axes is also expected to differ.

R Raff: In your scenario of evolutionary course, you showed a primitive butterfly with white wings and black spots. Have you looked to see whether these black spots show the same kind of genesis as eyespots genetically? They don't look like eyespots.

Brakefield: I agree; we should do this and we haven't. I don't know what the answer would be. If we can culture them this should be a straightforward observation to make.

Duboule: For a non-expert in butterflies and natural selection, I am a bit confused. You said that in some species the eyespots attract the predators to aim away from the body when they attack, while in other species the eyespot puts predators off.

Brakefield: Yes, both these potential functions can apply. In general, large, conspicuous eyespots that are flashed at predators are thought to have a startling type of function, whereas smaller, more marginally-located eyespots are considered to act as targets for attacks, deflecting them away from the vulnerable body. Both those functions may be parasitizing some resemblance to the vertebrate eye.

Duboule: What would be your argument against the following provocative statement. Distalless is downstream of the Dpp pathway. Likely it is also downstream of the Wnt signalling pathway. Perhaps there is a load of hedgehog and Wnt in the wing for doing something else and, for some intrinsic reasons, there is a leakage of the system and you see these eyespots as a collateral effect of the genuine function of Dpp and wingless in the wing, which means that they are not at all the target of natural selection; they are just there because they are selected for by another mechanism acting in the wing.

Brakefield: I think what you have done is to describe eloquently a potential scenario for the co-opting of those pathways into making eyespots. But I would combine this with the unequivocal evidence that these eyespots are very important in terms of fitness, survival and reproduction in the wild. The best experiment is that of Nico Reitsma performed in African forests in Malawi. These butterflies show striking phenotypic plasticity for the wing pattern on the ventral wings that are exposed when the adults are at rest on foliage or the forest floor. If the larvae are raised at low temperature the butterflies have no ventral eyespots, whereas marginal eyespots and a white medial band develop after exposure to higher temperatures. In the wild, there is a wet season form with large eyespots during the rainy season, followed by a single generation of the dry season form without eyespots that develops when the temperature is declining at the end of the rains. You can do beautiful experiments in the field. By using higher temperatures, wet season form butterflies can be raised in the lab during the dry season. These can then be released as marked butterflies into the forest. Mark–release–recapture experiments have demonstrated that these butterflies have no chance of surviving the whole of the dry season because of higher predation. You can then also take wild dry season butterflies from another forest and paint artificial eyespots on to their wings. This takes about five minutes a butterfly. These artificially painted butterflies can then be released with controls who have inconspicious, brown eyespots painted on that aren't visible. The same happens: the ones with the painted on conspicious eyespots disappear rapidly. This is robust evidence that eyespots really matter in terms of survival.

Stern: No, that shows that they are really dangerous in the dry season!

Brakefield: If you do the same experiments in the wet season there are very small, but repeatable differences in fitness, but in the opposite direction. Thus there is an advantage in having the eyespots in the wet season although it is a small one; evidently the deflective function does work sometimes. In experiments in the lab, eyespots seem to matter only for naïve birds and not for lizards.

Moving back a little, Denis Duboule, I love your scenario: something like this evidently happened in terms of co-opting those signalling pathways to do something completely different to produce one of these serial patterns of wing elements. However, I disagree strongly with the idea that the eyespots don't matter in terms of natural selection. Mind you, what sort of influence natural selection played in the early stages when the novelty may have involved the tiny, undifferentiated spots along the wing borders in primitive Lepidoptera can only be a just-so story. Perhaps they were just part of a cryptic pattern on certain types of resting backgrounds.

Coates: The other main predator on butterflies is natural historians. The effect of this is that anyone who has worked with museum collections knows that you get a distorted sample that emphasizes diversity rather than relative abundance.

Brakefield: As we speak there is a graduate student in South Africa who is collecting at several localities. She has already done this in Uganda and we hope that she will also go to Ethiopia. We will cover the whole latitudinal gradient of *Bicyclus anynana* and begin to look into the inter-population, spatial variation in these phenotypes within a species. At the same time we would like to take 'random' samples of as many species as possible. In a single forest in the Cameroon you can collect 40 of these *Bicyclus* species in traps within a few weeks.

Hall: You could go back to a museum collection that was collected 100 years ago and look at the variation.

Brakefield: There are indeed some interesting examples of precisely this type of use of museum collections.

Lieberman: Another selection experiment that relates to tinkering: you showed that some aspects are constrained and others aren't. You made the argument that selection is trying to mimic what an eye looks like. Can you then distinguish between the selective advantages and disadvantages of two colours? Is there any advantage to differences in colour? Is the constraint because there is no selective advantage, or because of some network pathway?

Brakefield: You can put this in all sorts of ways. The way I like to think about it is that perhaps there has been evolution for evolvability for one trait along that axis, and not for the other trait? This is the natural selection explanation for what is going on. I don't know the answer. The alternative is that the developmental properties of changing colour are much more difficult to evolve evolvability in, than size. These alternatives are interesting, and I think we will have the tools to examine them further and begin to make more valued judgements.

Wagner: Why is it difficult to change the colour-composition of different eyespots in different directions?

Brakefield: The more-or-less throwaway argument at this time is that the thresholds on which colour appears to depend are a wing-level response. There is a layer of epithelial cells across the whole wing. This appears to be the explanation for why all the eyespots have a similar colour composition.

Wagner: The signalling centre is also part of the epithelium. This is regionally different. It is possible to have epithelial cells react differently with at least one pathway.

Brakefield: Colour isn't determined at the early part of the pathway; the responses are right at the end of that developmental pathway. It might then be a timing issue that constrains or biases the generation of variation.

R Raff: These butterflies aren't just static. So, in a sense, behaviour should vary. If the spots are meant as a lure, then the butterfly might be expected to hold its wings still, but if it is a startle thing then the spots should be kept hidden until the butterfly is disturbed, and then flashed.

Brakefield: Absolutely. Unlike *Bicyclus*, some related species of Madagascan butterflies (in the genus *Heteropsis*) that we have begun to work with do flick their wings open when disturbed. These wings have a more centrally sited eyespot with a very broad gold ring. This eyespot is, to our eyes, highly conspicuous when the butterflies flick their wings; they may also hold them open in an apparently ritualized type of display for a few seconds. The peacock butterfly in northern Europe will do something similar. Thus, the *Bicyclus* and *Heteropsis* genera do show this behavioural difference. It is almost certainly strongly associated with a difference in the way the eyespots are used against predators. The Madagascan butterflies have very small ventral wing eyespots; there is a dry season-like form throughout the year without any deflective eyespots. They seem to rely on crypsis, and if this fails they will try to intimidate the predator.

Budd: This idea of particular traits being decoupled from others in terms of selection is interesting. Has anyone done any classical phenocopy experiments to try to uncover mechanism of hidden variation behind this?

Brakefield: I should thank you for the question, but it is a minefield. In Fred Nijhout's book on butterfly wing patterns, there is a wonderful chapter in which he goes through the old literature. This sort of experiment was beloved by many early lepidopterists. They put pupae of various species in the fridge for a short period in early pupation. We have tried this in different ways with *Bicyclus* and see little or no effect. Effects observed in some of those earlier studies are extraordinary, and I find it interesting in relation to some of the topics we have been talking about regarding uncovering hidden genetic potential. In particular, there are some Californian Vanesid butterflies in which if certain temperature treatments are performed in early pupae, it is possible to produce close phenocopies of other species from the same lineage. Some of these patterns are apparently never shown by the particular species itself. Mechanistically, it is just a black box at the moment but it could be an interesting experimental system. In our system we see very subtle effects from heat shocks. Sean Carroll and Mike Serfas have been doing in some ways similar experiments using compounds such as lithium chloride known to interfere with (candidate) developmental pathways. These are injected during early wing development in the pupal stage. Sometimes profound effects are seen on the wing patterns which are reminiscent of some of those produced by cold or heat shock experiments

Hallgrimsson: When you transplanted the central foci of the spots, the responses you got were diffuse. Was this experimental noise, or is there something about the local environment that makes the spot develop in this way?

Brakefield: I presume that it is some sort of composite explanation. Basically we are producing a major wound on the wing, which results in a lot of wound-healing around the grafted tissue before the signal can be propagated. There could also be many other components of developmental noise. We seldom observe a sharp boundary between colour rings in the ectopic grafted eyespots

Bard: Looking through the programme of this meeting, it seems to me that the most obvious cause of tinkering is not being discussed. This is modifying the activity of an individual molecule through mutation. I have no experimental data in mind, but an obvious theoretical example is in the production and interpretation of morphogen patterns. If there is tinkering with the molecules that 'read' morphogen concentrations, then the resulting pattern will be altered. A nice example comes from considering simple 2D Turing patterns where the kinetics generate complementary patterns of concentration peaks surrounded by troughs and concentration dips within a higher concentration field. Merely by tinkering with the abilities of the molecules that interpret such concentrations, it is possible to generate patterns sufficiently different that, were they to generate hair patterns on mammals, they would be recognized as different varieties (Fig. 1 [*Bard*]). A similar situation occurs in simulations of butterfly patterns where small changes in the patterning parameters can generate very different patterns. Perhaps the most interesting work here has been done by Barios and his colleagues in Mexico (e.g. Barrio et al 1999). Simple tinkering with reaction-kinetics rate constants can generate quite remarkable classes of patterns.

Lieberman: Along those lines, have there been reaction-norm type experiments with the signal? Is there a threshold response, or a non-linear response to the signal?

Brakefield: This hasn't been done experimentally, but in terms of modelling, many people have addressed this. I think these models are exciting, but I would love to know how we could test some of these ideas experimentally.

Bard: People who make models have a moral obligation to suggest experiments for testing their models—unfortunately, standards of morality are low here!

Weiss: Within species you have an shown ellipse of phenotypes. Are they inbred?

FIG. 1. *(Bard)* Two versions of the same Turing pattern (a,b), but with different concentration thresholds: the patterns map to the Masai (c) and Cape (d) giraffes (from Bard 1981).

Brakefield: No, they are highly outcrossed. Actually, we can't easily inbreed them because egg hatching is extremely sensitive to inbreeding; there is a very high genetic load.

Weiss: You could be getting stochastic variation. The ellipse of phenotypes in a population from which you do your selection experiments could be stochastic variation or it could be modification of other genes.

Brakefield: The genetic correlation for a particular trait such as size or colour among a pair of eyespots is strongly positive and higher than 0.5. However, the genetic correlation between these two traits is very low. We haven't touched on what maintains this genetic variation in natural populations, but this is an equally interesting issue. Variation in space (e.g. along altitudinal gradients) and in time (e.g. the dry seasons vary in severity) in combination with some adult movement probably contributes to this process.

Bell: Another way to interpret that variation might be as follows. You have a lot of variation in eyespot size but not colour.

Brakefield: No, along the axis of coupling among eyespots there is a more or less equivalent phenotypic variation for colour, and while you can't change it away from that axis you can change it along the axis. The genetic correlations are also similar.

Hall: An interpretation you gave during your talk was of the strength of the morphogen that was coming from the focus. If this morphogen were something like *Distalless*, you could see whether this was strength of molecules per cell, or whether you are getting more cells producing the same amount of morphogen.

Brakefield: There are more cells. Larger eyespots are associated with more of the central, organising cells, assuming that the focus translates more or less into the white pupil of the eyespot that is visible on the adult wing. If this is the case, there is probably quite a tight correlation between the number of cells and focus properties such as source strength or activity. However, we have no hard data and experiments in which we have attempted to transplant only part of, rather than a whole focus do not work, presumably because of the effects of damage.

Duboule: Distalless is a transcription factor. It is difficult to reconcile this with the orthodox definition of a morphogen.

Brakefield: Absolutely, *Distalless* encodes a transcription factor. It is one of the developmental genes up-regulated early on in eyespot morphogenesis but it is not a morphogen.

Bell: Lest there be any lingering doubt that these spots are selected by predators, there are similar spots on fish. There is experimental evidence that a spot on the tail of a fish will deflect the point of attack of a predatory fish towards the tail (McPhail 1977). There is further evidence pointing towards the importance of eyes and eye spots: lots of fish have cryptic markings that obliterate the image of the actual eye, for example by stripes across the eye. Eyes really matter to predators. There is a parasite that alters colour in stickleback, turning the eye into a big black object while at the same time turning the skin white. This parasite is manipulating this fish to make its eye more conspicuous and facilitate predation by birds because the parasite reproduces in the bird's gastrointestinal tract (LoBue & Bell 1994).

Brakefield: Eyespots don't seem to matter for lizards. Using experiments in which butterflies with manipulated eyespot patterns are presented as prey to captive *Anolis* lizards, we can find no evidence that the attacks of even naïve lizards are deflected by eye spots. If you walk through one of the East African woodlands during the dry season, the forest floor is rich in small lizard species, so we used to think that lizards were very important in terms of eyespot evolution. Now we don't think so. In South America, where *Anolis* lizards are found, there are other genera of butterflies which are doing everything that these African *Bicyclus* butterflies do in terms of phenotypic plasticity and eyespot evolution. Our results to date suggest that birds are more likely candidates for predators that make an impact on eyespot evolution.

Hall: Have experiments been done that look at the amount of damage a wing can sustain in relation to eyespot position?

Brakefield: Yes, but the problem with interpreting all these experiments is that we are looking only at the butterflies that have escaped. It is difficult to analyse the success or non-success of these different butterfly pattern elements.

Duboule: What is the life cycle for these butterflies? How long does it take before they mate?

Brakefield: The wet season butterflies which emerge as adults in a favourable environment in terms of temperature and food, mate within one or two days and lay eggs over the following couple of weeks or so. Dry season butterflies emerge at the beginning of the dry season and must then survive for seven or eight months before they can mate and reproduce at the beginning of the next wet season. Thus, there are two very different strategies for coping with the seasonality in their environment.

References

Bard JBL 1981 A model for generating aspects of zebra and other mammalian coat patterns. J Theoret Biol 93:363–385

Barrio RA, Varea C, Aragón JL, Maini PK 1999 A 2D numerical study of spatial pattern formation in interacting Turing systems. Bull Math Biol 61:483–505

LoBue CP, Bell MA 1993 Phenotypic manipulation by the cestode parasite *Schistocephalus solidus* of its intermediate host, *Gasterosteus aculeatus*, the threespine stickleback. Am Nat 142:725–735

McPhail JD 1977 A possible function of the caudal spot in characid fishes. Can J Zool 55:1063–1066

GENERAL DISCUSSION I

Brakefield: The more I think about the term 'tinkering' the more I think of it as being not the early co-option or source of key innovation or novelty early on, and perhaps not even the elaboration stage when the phenotype becomes more complex. Instead, I think of everything thereafter. But this doesn't seem to be the consensus view. I think it must depend on the scale of one's perspective about evolution, and at what scale one's own particular interest lie. From what I have heard at this meeting, most people think of tinkering as including more or less everything in evolution. To me, the word suggests that what is occurring is that two species in subtly different environments take advantage of the evolvability of traits to adapt, each to their own environment.

Weiss: Isn't that's what is always going on? Isn't that evolution? At any given time and place that is what is happening.

Brakefield: It comes back to what is an innovation or a novelty or a co-option.

Lieberman: I suspect that coming up with an absolute definition of tinkering that everyone would agree on is a silly, futile exercise. We could come to a consensus definition, though.

Cheverud: Big changes start out as small changes in evolution that are later built upon.

Lieberman: Changes within a body plan are probably what most people would construe as tinkering. Changes between body plans wouldn't fit into Jacob's analogy.

Cheverud: That doesn't really happen in that at the time they are formed, body plans are variable and not recognizable as such except in retrospect.

Bard: Tinkering implies, by definition, that we must have something to tinker with.

R Raff: The within body plan definition is not reading it right. If we run down a phylogeny way down deep, we find something as a precursor of a body plan; it is a less derived organism. It is not like trying to convert a vertebrate into an insect, or as Patten argued in his enormous 1912 book where he tries to convert a horseshoe crab into an armoured fish by turning it over and moving the mouth. The events that produce these later plans have to have occurred very early and involved events that were pretty minor at the time.

Wilkins: And involved tinkering.

Lieberman: At some level, all evolution has to occur by tinkering: think of the mammal-like reptiles and the reptile-like mammals.

R Raff: The events are not necessarily equivalent. It is likely that as metazoans evolved, constraints and modules get built in as features are added. These things themselves get canalized as the organism becomes more complex in its development.

Weiss: If we were all prevertebrates, at some time, 500–600 million years ago we would probably have seen canalization among the things that were existing at that point. It wouldn't be as complex, but still, the phenomena and processes would be relevant. There is currently a heated debate about horizontal transfer: the main idea is that this becomes impossible very early in cellular evolution because cells get committed to networked pathways and can't import something from other species.

Budd: It is an easy hypothesis to say that since the beginning of animal time development was somehow less canalized and things have hardened off since then, but it is difficult to test what this idea might mean. When we look at Cambrian animals it is hard to say that they appear less canalized than animals around today.

Stern: Let's call them less variable, then. There are data relevant to this. One study is by Amy McCune (1990) who studied fossil lake beds. She was able to look at the fish population present each year in stratigraphic slices from ancient lake beds. She found that as the original population expanded, there was a huge increase in the diversity of skeletal morphology. Over time, these would get wound down to pretty much the same morphology. This is similar to work done by E.B. Ford and H.D. Ford in the 1930s on the butterfly *Melitaea aurinia*, where populations that were expanding would show lots of diversity, and as the population stabilized this would be whittled away. Why can't we think about the Cambrian explosion in the same terms? There was a huge new variety of environments opened up, resulting in the generation of diversity. These persisted because there were many niches available.

Coates: Relative to your frame of reference, in terms of the fish taxa you have just described, the huge diversity relates to variation in the fulcral scales along the dorsal midline. This would probably induce yawns among non-specialists!

Bell: In your examples you gave two scales of time that are dangerous to mix together. Population expansions in a single species presumably reflect an excess of resources, whereas expansions involving numbers of species over millions of years are constrained by other factors.

Stern: What is the evidence that would reject the hypothesis that the Cambrian explosion reflects simply a huge ecological opportunity?

Coates: We can only do comparative work. We need to look at taxa coming from the same node. Even so, this is hemmed around all sorts of inbuilt suppositions: if we model equivalence, what do these two taxa do?

Lieberman: There is a set of people here that think about tinkering from a developmental perspective, versus those who think of tinkering in terms of pattern. Is this a false dichotomy?

Bard: It reflects our developmental history! In practice: virtually all tinkering occurs during development—and it seems to have been Waddington who first put the emphasis here.

Hanken: One way to make headway might be to divide the problem into more discrete units. The work that Paul Brakefield talked about concerns variation within a genus. How do you accomplish variation among species within a genus? No one is talking about this variation being involved in the origin of higher taxa; we are restricting the focus. Generally, when people use the term tinkering they are including a certain component of adaptive radiation. They are taking a particular kind of organism and making more species to flesh out the tree of life. If one can agree that this is the nature of tinkering and how it occurs, then one can ask whether these kinds of changes have the potential to lead to more significant evolutionary innovations.

Bell: At the beginning of this meeting tinkering was defined as 'the opportunistic use of what is on hand, small scale, constrained and often purposeless'. This is a set of criteria that don't necessarily bear any relationship to one another. Some people think small scale, and that if it is large scale it is not tinkering. I like the first component, opportunism; stuff that comes along. To me, this is the essence of tinkering. Then I am interested in what controls what comes along, such as mutation or developmental constraints that prevent a population from evolving to occupy some area of the morphospace.

Lieberman: That was Jacob's definition, not mine!

Bell: It is a multifaceted definition. How we react to it depends on what facet appeals to us.

Morriss-Kay: I wonder about the term 'purposeless'. It suggests some kind of mind behind it, because it implies there are cases where there is change that does have a purpose behind it. Perhaps 'functionless' would be more appropriate.

Laubichler: Perhaps we should remember the first two and a half pages of Jacob's paper were about the worldview of science, contrasting science to religious and mythical explanations. We are getting into a particular context of the argument here, arguing for the advantage of science to be specific, focusing on individual instances and then identifying that as a potential problem of the loss of meaning. If you only describe increasingly specific situations, how do you gain knowledge? Out of this he says that tinkering is a concept that, while it happens at any different level from molecular to evolution, might serve to give a unifying or integrative perspective. We have to be aware of this kind of background for this paper before we get lost in all kinds of details here.

Morriss-Kay: I want to raise another word which has been used here. I come from a background of developmental biology and I haven't got a feel for the way people are using 'canalize'. In the introduction, the following question was posed: 'How much does development canalize and/or buffer organisms from saltational changes in development?' Can someone explain it?

Hall: None of us have used it in the way you used it earlier to describe the formation of a canal in the centre of a rod.

Morriss-Kay: That's true, so what it the alternative meaning?

Hall: I use it to mean that development is channelled in particular pathways which it is difficult to get out of.

Bell: And the buffering is against genetic or environmental perturbations.

Weiss: This assumes the aspects of networks that Adam Wilkins was talking about. When you talk about 'buffer' you are talking about redundancy or feedback or something else that resists change.

Cheverud: It is structured variation, instead of variation being randomly spread in all directions in equal amounts. It implies some dimensions vary more and some vary less with mutation.

Wagner: If you have canalization it also sets up thresholds. It is the prerequisite for discontinuous variation. Canalization and discontinuous variation are two different sides of the same coin.

Wilkins: In this sense, canalization is being used as a synonym for 'developmental constraint'. Most of us would rather use this term to describe a basic property that development has, not something that it does in evolution.

Weiss: If the trait is 'canalized' it means there is no viable mutation that can change it.

Wilkins: It is a resistance to being changed; it is not an absolute state.

Weiss: Isn't this just using different words for the same non-specific concept?

Wilkins: A 'developmental constraint' is something that prevents evolution from going in certain path, whereas a state of canalization is something inherent to the developmental process.

Weiss: That sounds like chasing the same word around through the dictionary. Operationally it means that there is no viable genetic way out.

Hallgrimsson: Canalization is one of these concepts where we often confuse pattern and process. It is usually used in reference to a pattern (reduction of variation). The processes that lead to a reduction of variation and thus produce canalisation are usually unknown.

Bard: That meaning is different from the way that developmental biologists mainly use the term, and their usage is Waddington's. He invented the word 'canalization' in the context of his metaphor of the epigenetic landscape which was itself based on the developmental trajectories that tissues undergo over time, and their modifications through mutation (Fig. 1a *[Bard]*). The landscape was

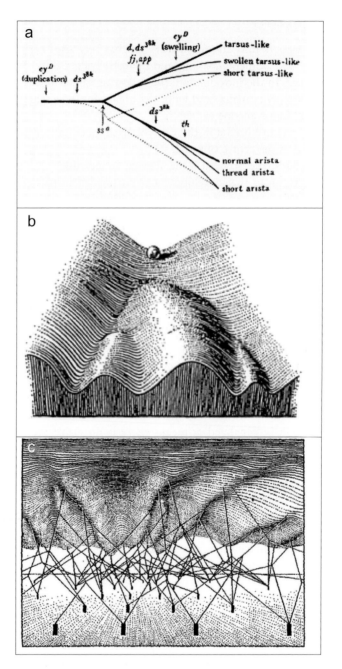

FIG. 1.   *[Bard]* (a) The normal trajectory of the *Drosophila* antenna imaginal disc development under the control of genes such as wild-type aristopedia (ssa) (lower branch) can be deflected into a limb (upper branch) through an ssa mutation. Mutations in other genes can further modulate the trajectories. (From Waddington 1940.) (b) The epigenetic landscape is a pictorial representation of how the fate of a tissue becomes fixed and stable as developmental time proceeds (From Waddington 1940). (c) A later drawing of the epigenetic landscape showing how its every feature is underpinned by (pegged to) the sets of underlying genes—the fate of tissues depends on the system as a whole, not a single gene. (From Waddington 1957)

essentially a mechanical model (Fig. 1b *[Bard]*) where development proceeded from high to low states as if driven by gravity. The landscape started with a high plane, off which were valleys with walls. The development of a cell was represented by the movement of a ball down the valleys, with the walls constraining movement downhill. In this landscape:

1. Development was *canalized* to follow the bottoms of the valley by these walls.
2. Stable pathways were called *creods*.
3. The increasing height of the walls above the valley bottom ensured the increasing stability of differentiation with time.

Most interesting, for its time, was the fact that the later version of this landscape was based on the behaviour of genes (the ties to the pegs in Fig. 1c *[Bard]*).

Hall: I think Waddington started with cells because he modelled the epigenetic landscape, with cells having only one or two options. Then this was translated over to the underlying genes.

Bard: Can I pull this back to Adam Wilkins' genetic networks of development? The current situation is that we can construct complex networks on the basis of expression and mutation, but we still lack the techniques, both analytic and numeric, for working out their properties. We will however have a nice assay: any theory that fails to generate the buffering (creodic) stability that characterizes developmental process will be wrong!

R Raff: 'Developmental constraint' is an evolutionary idea, not a developmental one. There the idea is that somehow the pre-existing system has possibilities that may be favoured and those which may not be. The word 'constraint' sounds like it is negative, but it doesn't necessarily mean that. It is unlikely that vertebrates will evolve wings like angels, not because it is impossible but because when wings have evolved there are already structures in existence that can be modified.

Weiss: We were assured earlier by Denis Duboule that we can't have six limbs. This rules angels' wings out.

Duboule: I don't think you could have limbs on your back!

References

McCune AR 1990 Evolutionary novelty and atavism in the Semionotis complex: relaxed selection during colonization of an expanding lake. Evolution 44:71–85
Waddington CH 1940 Organisers and genes. Cambridge University Press, Cambridge
Waddington CH 1957 The strategy of the genes. Allen & Unwin, London

Tinkering with transcription factor proteins: the role of transcription factor adaptation in developmental evolution

Günter P. Wagner and Anna Marie Pyle*

*Department of Ecology and Evolutionary Biology and *Department of Molecular Biophysics and Biochemistry, Yale University, Howard Hughes Medical Institute, New Haven, CT 06520-8106, USA*

Abstract. Evolution of transcriptional regulation is often seen as being driven by the origin of transcription factor binding sites in *cis*-regulatory DNA. But there is strong evidence that transcriptional specificity of transcription factor proteins is determined by protein–protein interaction with other transcription factors. Here we summarize the evidence that transcription factor protein function itself is evolving and suggest that this might play an integral if not leading role in the evolution of gene regulation.

2007 Tinkering: the microevolution of development. Wiley, Chichester (Novartis Foundation Symposium 284) p 116–129

Embryonic development is controlled through the regulation of gene expression (Carrolla et al 2001, Wilkins 2002). Consequently the evolution of development can be conceptualized as the evolution of differential gene regulation. This statement is true even if one acknowledges the importance of environmental and epigenetic factors in development and developmental evolution (Müller & Newman 1999, West-Eberhard 2003) since the genetic assimilation of a novel structure will also have to be inscribed into the genome to effect evolutionary change. In this short paper we follow this logic and ask what kind of molecular changes are involved in the evolution of gene regulation. The answer, of course, depends on the molecular interactions through which transcription factors regulate the expression of a target gene.

Gene expression is a complicated process that includes multiple levels and stages of molecular interactions including the nuclear localization of a transcription factor, chromatin modification, the binding of the transcription factor to DNA, the interaction with other transcription factors and non-coding RNAs and with the basal transcriptional complex, the processing and editing of the primary transcript, the stability of the mRNA or its degradation due to miRNAs and the translation into protein and the cellular localization and secondary modification of the protein, etc. Any of these steps in gene expression can be the target of regulation, but usually the greatest emphasis is put on transcriptional regulation, i.e. how transcription factors influence transcriptional initiation (e.g. Davidson 2001). Whether this emphasis on transcriptional regulation is ultimately justified is hard to judge, but the evolution of transcriptional regulation certainly plays a part in developmental evolution, and we also will focus here on transcriptional regulation.

Transcription factors interact with a target gene by binding to DNA and/or other transcription factors (Davidson 2001). Most transcription factors have a DNA-binding domain, in fact that is part of their definition, but there is a growing consensus that DNA binding is not specific enough to explain functional specificity (Grienenberger et al 2003, Mann & Morata 2000). The best supported model to explain functional specificity assumes that binding to certain DNA sequences is enhanced via protein–protein interactions with other transcription factors (Ryoo & Mann 1999) and through interactions with non-coding RNA. There are important cases where transcription factors have no DNA-binding capability. In fact, there are examples where a homeodomain protein, i.e. a protein able to bind to DNA, does not necessarily bind to DNA in its active state *in vivo* (Prochiantz & Joliot 2003, Topisirovic et al 2005). Strictly speaking, in this situation it is acting as a co-factor. Furthermore the binding to DNA target sequences can also be mediated through small RNA molecules, rather than a direct protein–DNA contact (Feng et al 2006, Sanchez-Elsner et al 2006). Hence a multitude of molecular interactions, including DNA, proteins and RNA, is involved in transcription factor function.

In evolutionary developmental biology, however, the evolution of transcriptional regulation has been seen as a problem of the evolution of the transcription factor binding sites (Wray et al 2003). The rationale for this perspective in part, is that transcription factors have been demonstrated to remain functionally equivalent over vast phylogenetic distances. For example, *Drosophila* Deformed protein can be replaced in some functional contexts by its vertebrate semi-orthologue, HoxD4 (McGinnis et al 1990). Distant paralogues, like HoxA3 and HoxD3, can replace each other in some functional contexts (Greer et al 2000), and even the homeodomain of HoxA13 is equivalent to that of HoxA11, which are even more distantly related than HoxA3 and HoxD3, with respect to body axis, kidney and male reproductive functions (Zhao & Potter 2001).

This conservation of transcription factor function, however, is not universal (see below) which raises the question what role the evolution of transcription factor proteins may play in the evolution of gene regulation. Furthermore, it has been argued that these types of factor swapping experiments overestimate the extent of functional conservation (Hsia & McGinnis 2003). It is known that in some cases 'complex' phenotypes are induced by a transcription factor by regulating only one or very few target genes. For instance, the celebrated case of *Pax6*, the mouse gene able to induce ectopic eye development in *Drosophila* (Halder et al 1995), is due to its influence on only two downstream genes (Chen et al 1997). Consequently anything affecting the function of these two genes will have the same effect as *Pax6*.

In the next sections we review evidence from various systems and approaches that show that transcription factor protein evolution is an important feature in developmental evolution. Evidence for the fact that transcription factors do not remain functionally equivalent in evolution comes from two main sources. On the one hand there is experimental evidence from developmental biology based on knockdown and knockout rescue paradigms or mis-expression experiments. On the other hand there is evidence from molecular evolution showing that transcription factor genes frequently undergo modification by directional selection.

Evidence from developmental biology

The classical approach to test for functional equivalence of transcription factors is to express a homologous gene from another species in a genetic model organism and record its regulatory effect. This was first done by McGinnis et al (1990) with a human semi orthologue of the *Drosophila* homeotic gene *Dfd*, *HoxD4*. The authors expressed the human *HoxD4* via a heat shock promoter in the *Drosophila* larva and showed that the transiently expressed human gene activates the transcription of the endogenous *Dfd* gene. Auto-activation of *Dfd* is a specific regulatory function of *Dfd* in *Drosophila* development and occurs, restricted to certain stages of development, only in the ectoderm. The same, though less efficient, regulatory effect has been demonstrated for human *HoxD4* in *Drosophila*. This was the first evidence that the function of the paralogue group Hox4 proteins remained conserved since the most recent common ancestor of mammals and insects.

It was found, however, that this conservation does not extend to all functions of a transcription factor. To our knowledge, the first instance where this was demonstrated was a study comparing the regulatory activities of the *Drosophila tinman* gene and its vertebrate homologue *Nkx2.5* (Ranganayakulu et al 1998). Both genes are involved in heart development in their respective species. Ranganayakulu and collaborators used a rescue approach in *tinman* null mutant line of *Drosophila*. Injection of *tinman* mRNA into the *Drosophila* embryo restores the expression of the *tinman* target genes *FascIII, eve, ζ fh-1* and *D-MEF2*. The injection of human Nkx2-5 mRNA can restore the expression of *FascIII* but not of *eve, ζ fh-1*, and *D-MEF2*, clearly showing differences in the regulatory activity of the insect and mammal *tinman* homologues.

A particularly well investigated case of functional non-equivalence is that of *Ubx* in *Drosophila* compared to its homologues in velvetworms (Galant & Carroll 2002, Grenier & Carroll 2000) and brine shrimp (Ronshaugen et al 2002). Grenier and Carroll (2000) compared the in vivo activity of velvet worm *Ubx* (*Acanthokara kaputensis*; Onychophora: *O-Ubx*) with that of the Drosophila *Ubx1a* (*D-Ubx1a*) in a mis-expression approach. O-Ubx and D-Ubx1a differ in their homeodomains by only two substitutions, but their overall similarity is low. O-Ubx is only 214 amino acid long (AAB92412) compared to D-Ubx isoform 1a of 380 amino acids (AAA84408). Like D-Ubx, O-Ubx can transform antenna towards a leg pheno-type and forewing into a haltere. More specifically O-Ubx can repress the SRF gene in the wing disc and drive the expression of *dpp* in the visceral mesoderm, a specific target of D-Ubx. But other typical effects of D-Ubx can not be reproduced by O-Ubx. For instance the transformation of thoracic cuticle into abdominal cuticle is a typical effect of ectopic D-Ubx expression, but O-Ubx is not able to cause this phenotype. Similarly, the repression of Dll by D-Ubx in the leg rudi-ments is also not affected by O-Ubx. These D-Ubx specific activities are caused by differences outside the homeodomain, as shown by the expression of chimeric proteins where the O-Ubx was replaced by D-Ubx homeodomain. The authors interpret the results as evidence that the D-Ubx protein is able to engage in specific protein–protein interactions that the O-Ubx is lacking. In a later paper the sequence responsible for the D-Ubx-specific activity was identified as a repressor domain C-terminal of the homeodomain, a $QAQAQK(A)_n$ motif (Galant & Carroll 2002). Similar results were obtained in a comparison of D-Ubx and *Artemia* Ubx (Af-Ubx) by McGinnis and collaborators (Ronshaugen et al 2002), with the difference that Ronshaugen identified the Dll repressor domain to be in the N-terminal region of the Ubx protein, and the QA motif to inhibit the repressive signal. A deletion of the QA motif in the *Drosophila* Ubx leads to few mild effects, reducing leg repres-sion by 20%. The latter result might either indicate that the QA motif is in fact not an active repressor of Dll expression or that there are extensive canalizing effects of the evolved *Drosophila* genomic background. The latter possibility was confirmed in a study in which the wild-type allele of Ubx was replaced with an allele that had the QA motif deleted (i.e. $Ubx^{\Delta QA}$) (Hittinger et al 2005). It was shown that QA has an additive (or better cumulative, since strict additivity was not tested) effect together with other peptide motifs (Tour et al 2005) and AbdA. Furthermore it was found that the QA motif has pleiotropic effects on other char-acters such as haltere development and bristle development on the third leg. This is interesting since the QA motif is conserved among insects but many of the characters affected by it are variable among insects. This shows that the peptide acquired additional functions since it originated.

That there are also specific functional differences between the homeodomains of paralogous Hox genes was demonstrated by the homeodomain swapping experiments of Zhao & Potter (2001, 2002). In their 2001 study Zhao and Potter replaced the *HoxA11* homeobox with the *HoxA13* homeobox in the mouse and found that the *HoxA11*$^{(a13Hd)}$ can rescue the *HoxA11*$^{-/-}$ phenotype in the body axis, kidney and male reproductive tract development. Interestingly the *HoxA11*$^{(a13Hd)}$ allele antagonizes normal HoxA11 function in limb development and causes homeotic transformation of the uterus epithelium into cervix epithelium. This is interesting because the *HoxA13* homeodomain did not change coincident with the evolution of the mammalian female reproductive tract. This may indicate that molecular interactions in which *HoxA11* is engaged during female reproductive tract development, which is a derived function in the mammalian lineage, require plesiomorphic amino acids in the homeodomain. The implication is that either the interactions of the HoxA11 homeodomain required for the female reproductive tract are ancestral, or that a novel interaction evolved but that the interaction partner adapted to the conserved HoxA11 homeodomain sequence.

These results were followed up with homeodomain swapping experiments between HoxA11 and HoxA10 and HoxA4. HoxA10 is even more closely related to HoxA11 than HoxA13 and the HoxA11$^{(a10Hd)}$ was functionally quasi-equivalent even in the female reproductive tract, in addition to the body axis, kidney and the male reproductive tract. In contrast, replacing the HoxA11 homeodomain with that of HoxA4 led to near recessive null function in all but the axial development. The pattern emerging from these studies is that even the homeodomain is functionally non-equivalent among paralogues with the more divergent paralogues being non-equivalent in more organs than the more closely related homeodomains of HoxA10 and HoxA13. It is also clear that functional non-equivalence is more likely for more derived functions (e.g. mammalian female reproductive tract) than for more ancient characters (e.g. body axis). That is, the older the paralogues and the more derived the characters the less likely the homeodomains are functionally equivalent. These results are interesting in the light of recent findings of adaptive evolution of the homeodomain following Hox gene duplication (Lynch et al 2006) and the fact that paralogue-specific amino acid residues in the homeodomain are likely to be engaged in protein–protein interactions (Sharkey et al 1997).

Perhaps the most radical changes in transcription factor protein function are those of *bcd*, *zen* and *ftz*, all developmental genes of *Drosophila* (see Hsia & McGinnis 2003 for a short review). Ftz is also the example where the relationship between protein function evolution and evolution of protein–protein interactions has been documented experimentally. In *Drosophila*, *ftz* has been shown to have two principal functions: CNS development and segmentation as a pair rule gene. CNS function seems to be old and conserved in all arthropods investigated. The segmentation function, however, is not even universal among insects as it is not found in grasshopper. There the gene has a Hox-gene-like homeotic function. The change from homeotic to segmentation function is associated with a loss of the Pbx interaction motif F/YPWM and the acquisition of the nuclear hormone receptor interaction motif LXXLL, required for the segmentation function (Lohr et al 2001).

The few examples reviewed above show that transcription factors are functionally modular and that different functionalities can evolve independently. Functionalities involved in shared ancestral functions are often conserved, but derived functions of one homologue are often associated with protein functionalities that are unique to one of the homologues. This raises the possibility that the evolution of novel functional roles of a transcription factor is associated with the evolution of novel protein functionalities. Of course, the comparison of transcription factor functions from a small number of taxa does not demonstrate that these functional differences are causally involved in the evolution of these novel functions, since functional differences can also arise by 'phenogenetic drift' (Weiss & Fullterton 2000), i.e. functional divergence after the origin of the novel function (Wagner et al 2000). To establish a causal role in the origin of evolutionary novelties a combination of molecular evolution and experimental methods is required. To our knowledge, this has not yet been achieved in any example. In the next section we summarize the evidence that directional selection, i.e. adaptation, is involved in shaping the evolution of transcription factor proteins.

Evidence from molecular evolution

Another type of evidence that demonstrates functional changes in transcription factor proteins are widespread and frequent comes from the statistical analysis of sequence evolution. The idea that developmental genes are highly conserved, predicted that the sequence evolution of the coding regions should be slow and clock-like. As soon as sequence information from various transcription factor genes and signalling genes became available, this prediction was found to be wrong (for a review see Purugganan 1998). It was found that the rate and pattern of sequence evolution is highly variable and mosaic, and in some cases associated with signs of adaptive evolution, providing evidence that the amino acid changes have beneficial functional effects. Sequence evolution of transcription factor genes is characterized by a mosaic of quickly evolving domains that are interspersed with highly conserved sequence blocks. The rapidly evolving sequence motifs, however, are functionally not dispensable (Kempin et al 1995). This result is consistent with the experimental evidence for functional modularity of transcription factor proteins reviewed above.

Accelerated rates of evolution of transcription factor genes has been associated with biologically significant processes such as domestication (Purugganan et al 2000), adaptive radiation (Barrier et al 2001), gene duplication (Hughes 1999) and morphological innovations (Lynch et al 2004). But the rate of sequence evolution is influenced by a variety of factors, including population size, gene duplication, mutation rate, selection and population structure. Hence it is essential to establish that the pattern of sequence evolution is specific to the focal gene and caused by directional selection, rather than by other factors.

In a study of sequence evolution of *HoxA11* it was found that the rate of non-synonymous substitutions is very similar among species as diverse as the coelacanth, the *Xenopus* frog and the chicken lineages (Chiu et al 2000). Furthermore, the rate of evolution was found to be accelerated in the mammalian lineage, which was intriguing because it was already known that this gene plays an essential role in female fertility. The gene is necessary for the differentiation of the endometrial stroma cells and is thus essential for the nidation of the embryo (Hsieh-Li et al 1995). This is certainly a derived function of the mammalian lineage raising the possibility that the HoxA11 protein experienced directional selection in connection with the acquisition of a new biological role. The prediction was confirmed using likelihood models and more extensive taxon sampling (Lynch et al 2004). To our knowledge this is the first case in which the acquisition of a novel biological role for a transcription factor is associated with directional selection on the amino acid sequence. Whether these amino acid substitutions are in fact necessary for endometrial cell differentiation is currently being investigated.

Another intriguing case of transcription factor protein evolution has been reported recently (Fondon & Garner 2005). It has been known for some time that transcription factor genes have a tendency to increase amino acid repeats, in particular in the amniotes and specifically in the mammals (Caburet et al 2004, Chiu et al 2000, Mortlock et al 2000). In some cases, increased amino acid repeat size has been shown to cause human congenital deformities (Karlin et al 2002). In a study of repeat length polymorphisms of 17 transcription factor genes among 92 dog breeds, Fondon and Garner found strong evidence that these repeats are under directional selection in the dog breeds and cause morphological differences. This shows that insertions and deletions of codons can contribute to rapid morphological divergence, in addition to the more conventional mechanism of amino acid substitutions and *cis*-regulatory changes. The molecular function of these amino acid repeats is not entirely clear, but there is some possibility that they provide opportunities for protein–protein interaction and may play a role in transcriptional repression, where alanine repeats are associated with repressor activity and polyglutamine repeats with transcriptional activation (Gerber et al 1994). The hypothesis about the alanine repeats, however, was not confirmed in the case of the alanine repeat of the mammalian HoxA11 protein. The alanine repeats do not contribute to repressor activity in an artificial recruitment assay (Roth et al 2005). Instead all repressor activity is mapped to the N-terminal part of the homeodomain. In the case of the dog breeds, however, it was found that a deletion of the PQ repeat of the *Alx-4* gene is associated with rear first digit polydactyly in the Great Pyrenees breed. The PQ repeat is known to be necessary for the binding of the Alx-4 protein to the lymphoid enhancer binding factor 1. These findings are intriguing because of the high rate of repeat mutations, which can provide genetic variation for morphological changes at a higher rate than point mutations.

Speculative coda

The molecular evolution evidence for the adaptive evolution of transcription factor genes in itself does not point to any specific biological function for which these proteins are selected. These changes could be related to their functional specificity in regulating gene expression, or could be due to unspecific effects on protein stability in different temperature and or intracellular regimes. The developmental evidence, however, clearly shows that some of the differences between orthologous and paralogous transcription factor proteins affect their functional specificity in regulating target genes. We speculate that the adaptive changes in transcription factor proteins are driven by selective forces related to their regulatory function. If this is indeed the case, it would follow that evolutionary changes in transcriptional regulation entail modifications among the protein–protein interactions between the transcription factors participating in the regulation of a target gene.

The idea that transcription factor protein evolution should play a role in adaptive evolution seems counterintuitive because changes to the transcription factor itself could have pleiotropic effects on the expression on all its target genes and in all the cell types it is expressed. If this would be the case, the chance that any of these changes are advantageous would be low and adaptive evolution of transcription factor proteins unlikely. This argument, however, overlooks the high degree of modularity of protein function, including the protein–protein interactions of transcription factors (Stern 2000). Detailed studies about the transcription factor protein complexes from different species and their sequence evolution are required to test this hypothesis.

References

Barrier M, Robichaux RH, Purugganan MD, Barrier M, Robichaux RH 2001 Accelerated regulatory gene evolution in an adaptive radiation. Proc Natl Acad Sci USA 98:10208–10213
Caburet S, Vaiman D, Veitia RA 2004 A genomic basis for the evolution of vertebrate transcription factors containing amino acid runs. Genetics 167:1813–1820
Carroll SB, Grenier JK, Weatherbee SD 2001 From DNA to diversity. Blackwell Science, Malden, MA
Chen R, Amoui M, Zhang Z, Mardon G 1997 Dachshund and eyes absent proteins form a complex and function synergistically to induce ectopic eye development in Drosophila. Cell 91:893–903

Chiu C-H, Nonaka D, Xue L, Amemiya CT, Wagner GP 2000 Evolution of Hoxa-11 in lineages phylogenetically positioned along the fin-limb transition. Mol Phylogenet Evol 17:305–316

Davidson E 2001 Genomic Regulatory Systems. Academic Press, San Diego.

Feng J, Bi C, Clark BS et al 2006 The Evf-2 noncoding RNA is transcribed from the Dlx-5/6 ultraconserved region and functions as a Dlx-2 transcriptional co-activator. Genes Dev 20:1470–1484

Fondon JW III, Garner HR 2005 Molecular origins of rapid and continuous morphological evolution. Proc Natl Acad Sci USA 101:18058–18063

Galant R, Carroll SB 2002 Evolution of a transcriptional repression domain in an insect Hox protein. Nature 415:910–913

Gerber HP, Seipel K, Georgiev O et al 1994 Transcriptional activation modulated by homo-polymeric glutamine and proline stretches. Science 263:808–811

Greer JM, Puetz J, Thomas KR, Capecchi MR 2000 Maintenance of functional equivalence during paralogous Hox gene evolution. Nature 403:661–665

Grenier JK, Carroll SB 2000 Functional evolution of the Ultrabithorax protein. Proc Natl Acad Sci USA 97:704–709

Grienenberger A, Merabet S, Manak J 2003 Tgfbeta signalling acts on a Hox response element to confer specificity and diversity to Hox protein function. Development 130: 5445–5455

Halder G, Callerts P, Gehring WJ 1995 Induction of ectopic eyes by targeted expression of the eyeless gene in Drosophila. Science 267:1788–1792

Hittinger CT, Stern DL, Carroll SB 2005 Pleiotropic functions of a conserved insect-specific Hox peptide motif. Development 132:5261–5270

Hsia CC, McGinnis W 2003 Evolution of transcription factor function. Curr Opin Genet Dev 13:199–206

Hsieh-Li H, Witte D, Weinstein M, Branford W, Li H 1995 Hoxa 11 structure, extensive antisense transcription, and function in male and female fertility. Development 121: 1373–1385

Hughes AL 1999 Adaptive evolution of genes and genomes. Oxford University Press, New York

Karlin S, Brocchieri L, Bergman A, Mrazek J, Gentles AJ 2002 Amino acid runs in eukaryotic proteomes and disease association. Proc Natl Acad Sci USA 99:333–338

Kempin SA, Savidge B, Yanofsky MF 1995 Molecular basis of the cauliflower phenotype in Arabidopsis. Science 267:522–525

Lohr U, Yussa M, Pick L 2001 Drosophila fishi tarazu: a gene on the border of homeotic func-tion. Curr Biol 11:1403–1412

Lynch V, Roth J, Takahashi K et al 2004 Adaptive evolution of HoxA-11 and HoxA-13 at the origin of the uterus in mammals. Proc R Soc B Biol Sci 271:2201–2207

Lynch VJ, Roth JJ, Wagner GP 2006 Adaptive evolution of Hox-gene homeodomains after cluster duplications. BMC Evol Biol 6:86

Mann RS, Morata G 2000 The developmental and molecular biology of genes that subdivide the body of drosophila. Annu Rev Cell Dev Biol 16:3861–3871

McGinnis N, Kuziora MA, McGinnis W 1990 Human Hox-4.2 and Drosophila deformed encode similar regulatory specificities in Drosophila embryos and larvae. Cell 63:969–976

Mortlock DP, Sateesh P, Innis JW 2000 Evolution of N-terminal sequences of vertebrate HOXA13 protein. Mamm Genome 11:151–158

Müller GB, Newman SA 1999 Generation, integration, autonomy: three steps in the evolution of homology. In: Homology. Wiley, Chichester (Novartis Found Symp 222) p 65–79

Prochiantz A, Joliot A 2003 Can transcription factors function as cell-cell signalling molecules? Nat Rev Mol Cell Biol 4:814–819

Purugganan MD 1998 The molecular evolution of development. Bioessays 20:700–711

Purugganan MD, Boyles AL, Suddith JI 2000 Variation and selection at the CAULIFLOWER floral homeotic gene accompanying the evolution of domesticated Brassica oleracea. Genetics 155:855–862

Ranganayakulu G, Elliott DA, Harvey RP, Olson EN 1998 Divergent roles for NK-2 class homeobox genes in cardionenesis in flies and mice. Development 125:3037–3048

Ronshaugen M, McGinnis N, McGinnis W 2002 Hox protein mutation and macroevolution of the insect body plan. Nature 415:914–917

Roth JJ Breittenbach M, Wagner GP 2005 Represson domain and nuclear localization signal of the murine Hoxa-11 protein are located in the homeodomain: no evidence for role of poly-alanine stretches in transcriptional repression. J Exp Zool B Mol Dev Evol 304B: 468–475

Ryoo HD, Mann RS 1999 The control of trunk Hox specificity and activity by Extradenticle. Genes Dev 13:1704–1716

Sanchez-Elsner T, Gou D, Kremmer E, Sauer F 2006 Noncoding RNAs of Trithorax response elements recruit Drosophila Ash1 to Ultrabithorax. Science 311:1118–1123

Sharkey M, Graba Y, Scott MP 1997 Hox genes in evolution: protein surfaces and paralog groups. Trends Genet 13:145–151

Stern DL 2000 Evolutionary developmental biology and the problem of variation. Evolution 54:1079–1091

Topisirovic I, Kentsis A, Perez JM et al 2005 Eukaryotic translation initiation factor 4E activity is modulated by HoxA9 at multiple levels. Mol Cell Biol 25:1100–1112

Tour E, Hittinger CT, McGinnis W 2005 Evolutionary conserved comains required for activation and repression functions of the Drosophila Hox protein Ultrabithorax. Development 132:5271–5281

Wagner GP, Chiu C-H, Laubichler M 2000 Developmental evolution as a mechanistic science: the inference from developmental mechanisms to evolutionary processes. Am Zool 40:819–831

Weiss KM, Fullterton SM 2000 Phenotypic drift and the evolution of genotype-phenotype relationships. Theor Pop Biol 57:187–195

West-Eberhard MJ 2003 Developmental plasticity and evolution. Oxford University Press, Oxford

Wilkins AS 2002 The evolution of developmental pathways. Sinauer Assoc., Sunderland, MA

Wray GA, Hahn MW, Abouheif E et al 2003 The evolution of transcriptional regulation in eukaryotes. Mol Biol Evol 20:1377–1419

Zhao Y, Potter SS 2001 Functional specificity of the Hoxa 13 homeobox. Development 128:3197–3207

Zhao Y, Potter SS 2002 Functional comparison of the Hoxa 4, Hoxa 10, and Hoxa 11 Homeoboxes. Dev Biol 244:21–36

DISCUSSION

Weiss: As I understand it, part of the classical view of the conservation of transcription factors is that they are used in multiple tissues, in multiple developmental stages. If you change the transcription factor, you potentially undermine all the traits. However, if just the binding sites are changed this can have just a local effect.

Wagner: That logic doesn't hold. There are tissue-specific and target gene-specific functional activities of the transcription factor protein itself. Not every change in the protein affects all functions of, let's say, the Hox protein. In the Tin/Nkx2.5 or Ubx examples, some of these activities are the same between, let's say, *Artemia* and *Drosophila*, but others like Distelless repression are not. The transcription factor protein itself is modular with respect to protein functions.

Weiss: But this does imply that the protein–protein part changes have to be compatible with all the different uses of the protein.

Wagner: You would expect that the conserved functions are important for all tissues, such as nuclear localization, transcriptional activity and DNA binding, are in a module by themselves in the homeodomain, and the tendency is that the more dynamic parts are encoded by the more variable part of the molecule. If you recall the alignment of Ubx, there is hardly any sequence similarity outside the homeodomain.

Duboule: The vast majority of homeodomains are encoded by a single exon, but there are exceptions to this rule such as for Hox14. Do these exceptions represent yet another way of tinkering? Does the splicing site correspond to the separation between functional domains?

Wagner: Actually, isn't the homeodomain of abdominal B also split in *Drosophila*? This is the homologue of all the 5′ genes. In vertebrates, all the Hox 9, 10, 11 and 13 homeoboxes are in one exon, and Hox14 that was lost in most of the vertebrates, but is present in shark and coelacanth and in these species the homeobox is still split. I think this is a secondary loss of an intron in the homeobox.

Duboule: Can this be used to separate functional domains?

Wagner: It is possible. As soon as you have an intron you can have differential splicing. But it would be strange. There is DNA binding and nuclear localization in the C-terminal part of the homeodomain. This isn't something you want to get rid of, but I don't think there are any co-factors that localize with Hox proteins.

Budd: Have you considered the dynamics of this process? Have you looked at the kinetics of these differences in protein–protein interactions?

Wagner: Before I got to this cartoon model I did a simple-minded mass action model. Essentially it boils down to the number of binding partners. Then it becomes a question of the product of the binding constants and the number of targets. The only remaining unknown was whether DNA binding is systematically stronger than protein–protein binding, but there doesn't seem to be evidence for this.

Cheverud: With the transcription factors, it seems that different mutations have different effects on pleiotropy. If you were to knock out a gene for a transcription factor you have these massive effects on many different characteristics. If, on the other hand, you have modifications of protein–protein interactions, these interactions could be specific to individual instances of gene expression, so you would have less pleiotropy.

R Raff: Do you have a semi-quantitative sense of whether most of what we see in tinkering in regulation of genes is going to lie in modifications like this of transcription factors, or in the response of downstream genes? Intuitively, it would seem that the latter would be the more common tinkering response, with *cis*-acting domains changing their binding strength or other properties.

Wagner: Binding sites themselves are highly conserved. There are weak ones and strong ones. From the perspective of evolutionary genetics, in population genetics we often ask ourselves what is the invasion probability of a mutation? There has to be a large enough fitness effect in the heterozygote state such that it can actually be 'seen' by natural selection. This logic seems to ask for a molecular mechanism as an initiating step that has a large enough biological effect to be seen by natural selection. If we latch onto an existing enhancer that already has lots of proteins, I would argue, that in many cases it would have a larger effect on transcriptional activity.

R Raff: In some cases, such as the axial display of Hox genes, we still see high concentration across taxa.

Wagner: Anteroposterior polarity and axis differentiation is an extremely ancient character. There you would expect protein function to be conserved, but this is not necessarily the case for more derived functions of the same transcription factor. They get recruited all over the place to fulfil new functions.

Hall: There are some genes that are privileged from tinkering because they have been around for so long, and they are so important that you can't tinker with them.

Wagner: The size of these proteins is changing radically. The deformed protein in *Drosophila* is 400 amino acids, but its homologues in mammals are 250. Others become larger and larger in mammals. The amino acids evolution can tinker with are also variable. You can make it more tinkerable.

Wilkins: I didn't understand why you placed such emphasis on the importance of the equivalence of binding strengths. Even if the protein–protein interactions are somewhat weaker, nevertheless they are frequent and effective for shorter time scales. They could be just as biologically relevant.

Wagner: Naively, I thought it could be that there is an order of magnitude difference between them that would eliminate the stochiometric advantage, i.e. that there are more interaction partners available.

Wilkins: They can't be that weak because we know that they are important for transcription.

Wagner: The easy argument is that they are occurring at the same molar ratio for the DNA interaction and the protein–protein interaction, so the affinities have to be equivalent.

Wilkins: A further point is that you are still emphasizing changes in transcriptional activity as a source of evolutionary change. Last year, Claudio Alonso and

I published an article (Alonso & Wilkins 2005) arguing that all the other levels of regulation are probably just as important in developmental evolution. We also argue that many of these changes are probably supplanted by transcriptional changes.

Wagner: I agree, except you have to decide to work on one part. In principle, I agree with you.

Stern: The protein–protein binding is presumably not dependent on a single amino acid. To change protein–protein binding associations may require multiple substitutions. Your equivalence of a single nucleotide change and a change in a transcription factor binding versus a single substitution to change nucleotide binding is probably not valid.

Wagner: The same is true for binding sites, right? For each type of interaction you need a small number of residues in the right place, and one mutation creates a sequence motif that can start interacting. In that respect I do not see a difference between protein–protein interaction motifs and transcription factor binding sites.

Stern: The target site for changes in *cis*-regulatory sites may be much larger than the number of possible mutations that would give changes in protein–protein interactions. There are other issues that may be relevant to the quantitative aspects. I buy your argument that perhaps the means of these two processes might be different for all the reasons you say. But it seems that what is most important is the distribution of effects, and in particular, the extreme values. How often do you get mutations that cause the largest phenotypic effects? These, as you argued, are the most likely to be picked up by natural selection. We have evidence that you can change some binding sites and have extreme changes in transcription. In addition, if those changes occur in the upstream regions of transcription factors themselves, then they can have lots of downstream effects. You can, of course, get large phenotypic changes through some mutations, just as you can get some large changes through changes in protein–protein interactions. The question is, what are the relative numbers of the kinds of extreme events you can get? These are probably the ones that are most important.

Wagner: We don't know at all. I don't mean to say that the binding site evolution itself is irrelevant. I am trying to oppose the exclusive attention that has been given to binding site evolution as the mode of creating new regulatory links. There is too much going on with the protein itself. This whole ideology of conserved transcription factors is a misreading of the empirical evidence. There needs to be some attention paid to the role of transcription factor protein evolution in that process.

Lieberman: I'm thinking of ways to test the hypothesis. If I understand you, you expect that as tinkering occurs you would have larger 'enhansomes'. Is there any way that other model organisms could be used to test this?

Wagner: The next thing we need to show is that the difference in activity between platypus HoxA11 and mouse HoxA11 is actually because they have a new binding partner. It is hard work but it is possible in principle. Once we have established enough evidence for function non-equivalence in this system, the next thing we'll do is look for a new protein interaction partner for HoxA11. Then the question is, is this partner a transcription factor that already was an existing ancestral enhancer? If the scenario turns out the way I expect then the platypus enhansome would be less complex than the placental one.

Bard: I have a biophysics question. The way you drew the binding site implied that there were so many surface interactions that the additive bit from the transcription factor binding to the DNA was almost irrelevant. There were five or six interactions and only one of them was on DNA.

Wagner: They still have to attach to DNA.

Bard: Are you saying that if they don't attach to the DNA they won't get to the rest?

Wagner: The binding activities are not additive; they bind co-operatively. The protein–protein interaction makes the protein–DNA association stronger.

Bard: How does this compare with kt? If thermal energy is not going to break up any of the wrong complex forms, you may just jam up your enhancersome with the wrong transcription factor.

Wagner: I didn't make any statements about the absolute value of the binding energy. Many aren't known. The argument I was making was a relative one based on independent evidence that the per angstrom squared binding energy in protein–DNA interactions is about the same as in protein–protein interactions. Therefore surface areas give us a proportional measure.

Hall: Might another way to test this be to look at situations where there is genetic redundancy? You knock out a Hox gene and there is one that can compensate, and another that can compensate but not quite as well? Could this be a way of getting at the way that tinkering is occurring at different levels?

Wagner: Yes, we could ask what proteins can be redundant. That's a good point. There is also an asymmetry in terms of a mixture of the proteins expected to evolve. Another potential derived protein interaction is between Gli-3 and HoxD12 in limb development. The part of the HoxD12 protein that interacts with Gli3 hasn't changed in evolution. This is probably because Gli3 is interacting with all 5′ HoxD proteins.

Reference

Alonso C, Wilkins A 2005 The molecular elements that underlie developmental evolution. Nat Rev Genet 6:709–715

Tinkering with constraints in the evolution of the vertebrate limb anterior–posterior polarity

Denis Duboule, Basile Tarchini[1], Jozsef Zàkàny and Marie Kmita[2]

Department of Zoology and Animal Biology, and National Research Centre 'Frontiers in Genetics', University of Geneva, Sciences III, Quai Ernest Ansermet 30, 1211 Geneva 4, Switzerland

Abstract. Genes belonging to both *HoxA* and *HoxD* clusters are required for proper vertebrate limb development. Mice lacking all, or parts of, *Hoxa* and *Hoxd* functions in forelimbs, as well as mice with a gain of function of these genes in the early limb bud, have helped us to understand functional and regulatory issues associated with these genes, such that, for example, the tight mechanistic interdependency that exists between the production of the limb and its anterior to posterior (AP) polarity. Our studies suggest that the evolutionary recruitment of *Hox* gene function into growing appendages was crucial to implement *hedgehog* signalling, subsequently leading to the distal extension of tetrapod appendages, with an already built-in AP polarity. We propose that this process results from the evolutionary co-option, in the developing limbs, of a particular regulatory mechanism (collinearity), which is necessary to pattern the developing trunk. This major regulatory constraint imposed a polarity to our limbs as the most parsimonious solution to grow appendages.

2007 Tinkering: the microevolution of development. Wiley, Chichester (Novartis Foundation Symposium 284) p 130–141

The development and evolution of tetrapod appendages provides a particularly enlightening example of what François Jacob quoted as 'tinkering' (Jacob 1977), or at least of one of the various interpretations one can give to this quotation. It is indeed understandable that, using this term, Jacob did not explicitly refer to a tinkering involving entire genetic pathways, but instead, the multiple usage of a core of building blocks and basic processes and their combinations, to generate biological diversity. This landmark paper was soon followed by a series of

---

Present addresses: [1]Department of Biology, McGill University, 1205 Ave. Docteur Penfield, Montréal, Quebec, Canada and [2]Unité de Génétique et Développement, Institut de Recherches Cliniques de Montréal (IRCM), 110 avenue des Pins Ouest, H2W 1R7, Montréal Quebec, Canada.

discoveries (e.g. the inter-species conservation of genes and networks), which not only revolutionized our views on developmental genetics, but also provided the conceptual tools to contemplate neo-Darwinism from a new perspective (see for example Duboule & Wilkins 1998, Kirschner & Gerhart 2006). In this novel context, the concept of 'tinkering' has survived well, to say the least, even though the associated notion of evolutionary 'genetic constraints' and the difficulty to reconcile this notion with an orthodox gradualist view of Darwinism still makes the impact of Jacob's prediction difficult to acknowledge for many of us.

In this respect, the historical and heuristic values of tetrapod limbs are of interest, as they represent the original example of structures that developed and evolved through the redeployment of a range of genetics pathways necessary for the ontogeny of the major body axis (Dollé et al 1989). Nowadays, the mere fact that limbs are rather recent innovations logically implies that they find their origin in the co-option of regulatory circuits that had previously evolved in a different context. While this is well accepted, the problems that it creates regarding the genetic constraints applied to the realm of available morphological possibilities, should not be overlooked. In this paper, we would like to discuss this issue and show that a large part of our basic limb morphology, hence its functionalities, is strongly constrained by those genetic mechanisms recruited to evolve these structures.

Hox genes, *Sonic hedgehog* (*Shh*) and limb development

Vertebrate limbs bud out of flank mesoderm through interactions with the overlaying ectoderm. The subsequent outgrowth and patterning of skeletal elements require signals from both the apical ectodermal ridge (AER) and the zone of polarizing activity (ZPA), a cohort of cells at the posterior margin of the bud, near the AER. These cells express *Sonic hedgehog* (*Shh*; Riddle et al 1993) whose product promotes distal limb growth and patterning, notably via its effect upon *Hox* genes belonging to both *HoxA* and *HoxD* clusters. Before responding to *Shh* signalling, several *Hox* genes are expressed in the early bud, some with a restriction for the posterior part where they may promote *Shh* transcription and/or maintenance (Zakany et al 2004).

Functional analyses have highlighted the role of *Hox* genes in developing limbs. In particular, compound mutants revealed synergistic and redundant mechanisms, as phenotypic alterations were significantly more severe than merely additive. While this raises problems in assigning gene specific phenotypes, it suggests that *Hox* products act quantitatively in both the production and organization of the structure. This conclusion is supported by the truncations observed in mice lacking either group 13, 11 or 10 *Hox* genes (Davis et al 1995, Fromental-Ramain et al 1996, Wellik & Capecchi 2003). In contrast to the *HoxA* and *HoxD* clusters, *HoxB* and *HoxC* are unlikely to play a major role in forelimb development (Wellik & Capecchi 2003, Nelson et al 1996), based on expression analyses. Furthermore, normal limbs developed when either of these gene clusters was deleted (Suemori & Noguchi 2000, Medina-Martinez et al 2000). Consequently, to evaluate the extent of forelimb development in the absence of any relevant *Hox* function, we engineered combined deficiencies of both the *HoxA* and *HoxD* clusters (Kmita et al 2005).

Because loss of *Hoxa13* is embryonic lethal, we floxed the *HoxA* cluster to generate tissue-specific deletions and used *Prx1-Cre* mice (Logan et al 2002), where *Cre*-mediated recombination occurs from early limb bud stage onwards. In this way, we produced mice lacking all *HoxA* and *HoxD* gene functions in forelimbs, which induced dramatic truncations of the appendages (Kmita et al 2005). At fetal stages, a single cartilage model was observed, articulating with the scapula. This cartilage element, bent in the middle, displayed a Y-like shape distally. We interpreted this as a truncated humerus, bent distally and followed by a bifurcation, which prefigured the formation of the zeugopod. Overall, mutant forelimbs appeared delayed in their development, as if patterning had been arrested at an early stage. Forelimbs lacking *Shh* also display severe distal and posterior agenesis, involving both the autopod and zeugopod, whereas the humerus is less affected (Chiang et al 2001, Kraus et al 2001). We looked at *Shh* expression and observed a virtually complete down-regulation in conditional *HoxD/HoxA* double mutant forelimbs, with only a few cells weakly positive. However, a single copy of either *HoxD* or *HoxA* was enough to trigger *Shh* transcription at a level similar to wild-type. To further investigate the requirement of *Hox* function for *Shh* transcription, we looked at early embryos deficient for both clusters obtained via *trans*-heterozygous crosses ($A^{-/-}; D^{-/-}$), before embryonic death had occurred. Two such embryos were obtained and *Shh* transcription was undetectable in the bud, whereas other sites showed normal expression levels in both cases (Kmita et al 2005). While these results indicated that the early expression of *Hox* genes in developing limbs is mandatory for *Shh* transcription to proceed, they paradoxically raised the question as to what restricts *Shh* transcription posteriorly, for several *Hox* genes are expressed throughout the early limb bud, including in most anterior cells where *Shh* is not normally activated.

Collinearity in limbs

In developing early limb buds, *Hox* genes are expressed following a collinear regulation whereby several 3′-located genes are expressed throughout the limb bud, whereas the transcription of more 5′-located genes (from *Hoxd10* onwards) is progressively restricted to more and more posterior cells. ZPA cells thus express distinct qualitative and quantitative combinations of *Hox* transcripts as compared to more anterior cells, and this may lead to the observed difference in *Shh* regulation. In support of this explanation, we analysed a stock of mice carrying a partial deletion of the *HoxD* cluster, leaving in place *Hoxd11*, *Hoxd12* and *Hoxd13*. Due to some regulatory re-allocations, these three genes were found expressed throughout the early bud, i.e. not only in posterior cells, as expected, but also in anterior cells from which their transcription is normally excluded (Zakany et al 2004).

In such mice, expression of *Hoxd11*, *Hoxd12* and *Hoxd13* in anterior limb bud cells induced the ectopic transcription of *Shh* anteriorly, leading to double posterior, mirror-image distal limbs (Zakany et al 2004). This result demonstrated that the ectopic expression of 'posterior' *Hox* genes in anterior limb bud cells was able to induce another ZPA. Therefore, it strongly suggested that the normal ZPA, in particular *Shh* transcription, is under the control of these 'posterior' *Hox* genes, which are normally only transcribed in posterior limb bud cells. From these experiments, it appears that the anterior–posterior (AP) polarity of the limb buds is partly fixed by the restricted expression of *Hox* genes posteriorly, which in collaboration with factors released by the overlying ectoderm (AER) trigger *Shh* expression in the posterior mesenchyme. Therefore, a key step in our understanding of this polarity is to uncover the mechanism that restricts *Hox* gene expression in posterior cells.

Hoxd genes are activated in limb buds following multiple collinear strategies. Early on, in the incipient limb buds, genes are activated in a time sequence starting with the most 3′ located members such as *Hoxd1* and *Hoxd3*. These genes are expressed throughout the emerging bud, with a rather homogeneous expression observed up to *Hoxd9*. Starting from *Hoxd10*, however, the expression domains become progressively restricted to successively more posterior limb cells, until *Hoxd12* and *Hoxd13*, as a set of nested patterns (Dollé et al 1989, Nelson et al 1996). Therefore, two collinear processes can be observed in the early limb bud, in time and space, the former hypothetically controlling the latter.

After early limb budding has occurred, once *Shh* is transcribed, a second wave of *Hoxd* gene collinear activation takes place, in the presumptive autopod domain, i.e. in those cells fated to generate the hands and feet. This expression of the most 5′-located *Hoxd* genes is controlled by sequences located far upstream of the cluster (GCR; Spitz et al 2003), following regulatory modalities that have began to be uncovered (Kmita et al 2002).

The mechanism(s) of collinearity

The collinear mechanism(s) underlying the first wave of *Hoxd* gene activation in limbs, in both time and space, was recently studied (Tarchini & Duboule 2006). A previous deletion of this gene cluster indicated that the main corresponding regulatory sequence(s) were localized outside the cluster itself (Spitz et al 2001). Subsequently, an engineered inversion of the same gene cluster revealed that the *Hoxd13* promoter, when placed at the position of *Hoxd1*, was expressed throughout the early limb bud, in a pattern related to this latter gene (Zakany et al 2004). These results indicated that the mechanism at work is promoter-independent and suggested that the progressive posterior restriction depends upon the mere position of a transcription unit within the cluster. They also led to the hypothesis that a critical element required for this collinear activation was located at the telomeric (3′) side of the cluster (ELCR; Zakany et al 2004).

To gain insights into this elusive early collinear mechanism, we produced and analysed of a set of mouse strains carrying a variety of deletions and duplications of parts of the *HoxD* cluster. These alleles were engineered using the targeted meiotic recombination strategy (TAMERE; Hérault et al 1998), starting with a selected set of parental lines such that breakpoints were readily comparable between various configurations. In these mice, gene topography was reorganized in many different ways, leading to important reallocations in their transcriptional controls during early limb budding. The analysis of such regulatory reallocations indicated that *Hoxd* gene collinearity in early limb buds is the result of two antagonistic regulations, implemented from either side of the cluster, which together establish the observed nested expression patterns in time and space (Tarchini & Duboule 2006).

The temporal aspect appears to be controlled by a sequence located telomeric to the *HoxD* cluster (ELCR), following a 'relative distance effect'. *Hoxd* genes located at the closest relative position (i.e. *Hoxd1*; *Hoxd3*) are activated first, whereas genes located at the other extremity of the cluster are activated last. This again seems to be promoter-independent and solely fixed by the genomic topography of a given gene. Therefore, the position of a given gene within the cluster will determine its timing of activation. However, the spatial collinearity is not solely determined by this timing process and also depends upon the existence of another, equally elusive, regulatory sequence located centromeric to the *HoxD* cluster, i.e. opposite to the ELCR. Here again, the relative distance between *Hoxd* genes and this sequence seems to be critical for the extent of posterior restriction (i.e. anterior suppression) in transcript distribution. Indeed *Hoxd* genes lying at the 5′ end of the series are strongly repressed in anterior limb bud cells whereas more 3′ located genes escape this repression and are transcribed throughout the limb bud (Tarchini & Duboule 2006).

Regulatory co-options

Regarding the evolutionary origins of these enhancer sequences, two alternative schemes—not exclusive from each other—can be considered. Firstly, a novel limb enhancer sequence may have emerged and been selected outside the cluster. Alternatively, pre-existing regulatory modules, positioned outside *HoxD*, may have been co-opted for yet another functional output in parallel with limb evolution. As far as the early phase of collinearity is concerned, it is likely that the two opposite regulatory influences derive from the second kind of scenario. Several aspects of this phenomenon are indeed reminiscent of the regulatory strategy implemented during the formation of the major body axis (the trunk), raising the possibility that part of this ancient trunk collinear regulation was recruited into the context of the newly growing limbs. In particular, the existence of two types of collinearities, temporal and spatial, which can be somehow disconnected from each other (reviewed in Kmita & Duboule 2003), suggests that the collinear strategy used during trunk development relies upon opposite mechanisms, much like the process described above for the early wave of activation in limbs. This is supported by the preliminary survey of the effects of our set of deletions/duplications upon the timing and location of *Hoxd* and *Evx2* gene expression in the developing trunk. A detailed analysis of this particular aspect will be informative in this respect and may shed light on this fundamental mechanism.

Therefore, we speculate that the early collinear activation in limb was recruited from the trunk mechanism, allowing for the distal growth of an ancestral appendage up to the wrist area. Subsequently, a second global regulation evolved (GCR; Spitz et al 2003), also located outside the cluster, which was necessary to accompany the emergence of the autopods (hands and feet). The existence of distinct regulatory processes for the two waves of *Hoxd* activation in limbs is coherent with the proposal that the proximal and distal parts of our limbs have different phylogenetic histories. In this context, it is noteworthy that the mechanisms resembling those implemented during the development of the trunk may control the early and proximal *Hoxd* gene expression, i.e. at a time and in locations where *Hox* genes are necessary to build the 'ancient' proximal part, whereas an apparently newly evolved enhancer accompanied the emergence of digits, i.e. of a rather recent evolutionary novelty. In this view, the various types of regulatory innovations, and their distinct mechanisms of co-option, may tell us about the phylogenetic history of the structure (Duboule & Wilkins 1998).

The limb AP polarity: recruitment of a regulatory constraint

One important effect of the early phase of collinear activation is the restriction of *Shh* signalling to the most posterior margin of the limb bud (Zakany et al 2004, Kmita et al 2005). Since *Shh* signaling is a major factor in the establishment of the limb AP polarity (Riddle et al 1993), this polarity appears to be the morphological translation of the asymmetry in the expression of some *Hox* genes, as a result of their early collinear expression. Consequently, the limb AP polarity may reflect nothing but a particular type of gene topography and its associated asymmetric regulations. Yet the major function of *Hox* genes in limb development is not to AP pattern the structure, but rather to trigger its growth, as the absence of *Hox* function leads to very severe truncations along the proximo-distal axis (Davis et al 1995, Kmita et al 2005). This apparent paradox suggests that the mechanism underlying the limb AP polarity did not evolve separately from, or in parallel with, the growth of the limbs. Instead, this mechanism was likely imposed as a collateral effect of the regulatory processes recruited to promote limb emergence and outgrowth.

In this view, an AP polarized limb is the expected consequence of using asymmetrically located enhancer sequences to control *Hox*-dependent outgrowth. Due to regulatory constraints imposed by the essential function of this gene family for trunk development, the co-option of this genetic system to promote limb development led to the impossibility to produce symmetrical limbs. During early trunk development, various combinations of *HOX* proteins are delivered at particular body levels, in specific cohorts of cells, which in turn will generate a given morphology. In tetrapods, axis formation and elongation are processes occurring along a time sequence; rostral structures are produced and determined before caudal structures. Therefore, it is crucial that the transcription of those *Hox* genes delivering 'caudal' information (e.g. *Hoxd13*) be postponed until the appropriate body level is produced, to prevent the premature formation of the caudal part of the body at a too rostral position. We believe this is the major evolutionary constraint maintaining temporal collinearity in vertebrates (Duboule 1994).

The co-option of this regulatory mechanism, along with the distal extension of appendages, transposed this repressive strategy into the context of growing limbs. As a result, 'caudal' *Hox* gene transcripts (e.g. *Hoxd11*, *Hoxd13*) are progressively restricted to the most posterior part of the emerging limb buds. Because these genes are able to activate and/or maintain the transcription of *Shh*, this later gene product became restricted to the posterior margin of the limb bud, hence generating an AP polarized structure. Therefore, our limb AP polarity reflects a major constraint that our body axis meets during its development and the regulatory solution that evolved to accommodate this constraint. In this scenario, the anterior to posterior polarity of our limbs merely results from the necessity to display caudal trunk structures at the extremity of our rostral to caudal major axis.

Acknowledgements

We thank F. Rijli, M. Logan and C. Tabin for their help with materials and collaborations, as well as all members of the Duboule laboratory and J. Deschamps for discussions. This work was supported by funds from the canton de Genève, the Claraz and Louis Jeantet foundations, the Swiss National Research Fund, the NCCR *'Frontiers in Genetics'* and the EU programmes *'Eumorphia'* and *'Cells into Organs'*.

References

Chiang C, Litingtung Y, Harris MP et al 2001 Manifestation of the limb prepattern: limb development in the absence of sonic hedgehog function. Dev Biol 236:421–435

Davis AP, Witte DP, Hsieh-Li HM, Potter SS, Capecchi MR 1995 Absence of radius and ulna in mice lacking hoxa-11 and hoxd-11. Nature 375:791–795

Dolle P, Izpisua-Belmonte JC, Falkenstein H, Renucci A, Duboule D 1989 Coordinate expression of the murine Hox-5 complex homoeobox-containing genes during limb pattern formation. Nature 342:767–772

Duboule D 1994 Temporal collinearity and the phylotypic progression: A basis for the stability of a vertebrate Bauplan and the evolution of morphologies through heterochrony. Development (suppl) 135–142

Duboule D, Wilkins AS 1998 The evolution of 'bricolage'. Trends Genet 14:54–59

Fromental-Ramain C, Warot X, Messadecq N, LeMeur M, Dolle P, Chambon P 1996 Hoxa-13 and Hoxd-13 play a crucial role in the patterning of the limb autopod. Development 122:2997–3011

Herault Y, Rassoulzadegan M, Cuzin F, Duboule D 1998 Engineering chromosomes in mice through targeted meiotic recombination (TAMERE). Nat Genet 20:381–384

Jacob F 1977 Evolution and tinkering. Science 196:1161–1166

Kirschner M, Gerhart J 2006 The plausibility of life. Yale University Press

Kmita M, Duboule D 2003 Organizing axes in time and space; 25 years of collinear tinkering. Science 301:331–333

Kmita M, Fraudeau N, Herault Y, Duboule D 2002 Serial deletions and duplications suggest a mechanism for the collinearity of Hoxd genes in limbs. Nature 420:145–150

Kmita M, Tarchini B, Zakany J, Logan M, Tabin CJ, Duboule D 2005 Early developmental arrest of mammalian limbs lacking HoxA/HoxD gene function. Nature 435:1113–1116

Kraus P, Fraidenraich D, Loomis CA 2001 Some distal limb structures develop in mice lacking Sonic hedgehog signaling. Mech Dev 100:45–58

Logan M, Martin JF, Nagy A, Lobe C, Olson EN, Tabin CJ 2002 Expression of Cre recombinase in the developing mouse limb bud driven by a Prxl enhancer. Genesis 33:77–80

Medina-Martinez O, Bradley A, Ramirez-Solis RA 2000 Large targeted deletion of Hoxb1-Hoxb9 produces a series of single-segment anterior homeotic transformations. Dev Biol 222:71–83

Nelson CE, Morgan BA, Burke AC 1996 Analysis of Hox gene expression in the chick limb bud. Development 122:1449–1466

Riddle RD, Johnson RL, Laufer E, Tabin C 1993 Sonic hedgehog mediates the polarizing activity of the ZPA. Cell 75:1401–1416

Spitz F, Gonzalez F, Peichel K, Vogt T, Duboule D, Zakany J 2001 Large scale transgenic and cluster deletion analysis of the HoxD complex separate an ancestral regulatory module from evolutionary innovations. Genes Dev 15:2209–2214

Spitz F, Gonzalez F, Duboule D 2003 A global control region defines a chromosomal regulatory landscape containing the HoxD cluster. Cell 113:405–417

Suemori H, Noguchi S 2000 Hox C cluster genes are dispensable for overall body plan of mouse embryonic development. Dev Biol 220:333–342

Tarchini B, Duboule D 2006 Control of HoxD genes collinearity during early limb development. Dev Cell 10:93–103

Wellik DM, Capecchi MR 2003 Hox10 and Hox11 genes are required to globally pattern the mammalian skeleton. Science 301:363–367

Zakany J, Kmita M, Duboule D 2004 A dual role for Hox genes in limb anterior-posterior asymmetry. Science 304:1669–1672

DISCUSSION

Brakefield: I can quite accept the idea that the vertebrate limbs are fundamentally evolved through *bricolage* setting up the potential, but I see no paradox in thinking that what you are looking at is a beautiful adaptive trait, but one whose origin is in the type of effects that you are talking about. Setting up the potential and capacity is part of evolvability.

Duboule: Absolutely. The best example for me to mention is the expression in the developing digit, because this is our own work. I can admit that we attributed a high adaptative value to this pattern over the past 10 years. Ultimately, it turns out that it is a rather stochastic process that simply generates a pattern. We contemplate this pattern and try to give it a sense it may not have.

Brakefield: Evolution is so opportunistic that it just makes use of this *bricolage* capacity for building, and ends up with something that is extraordinarily functionally efficient.

Carroll: The main subject of your paper was mice, which is an amniote, a mammal. But how do you explain how salamanders, and an extinct group of amphibians from the Palaeozoic, the branchiosaurs, develop their fore and hind limbs from anterior to posterior rather than the other way round, but retain the adult structure, with the same phalangeal sequence, as the amniotes. This is a well known conundrum in living salamanders. In fact, we have superb fossils from the carboniferous, 300 million years ago, which preserve developmental sequences from hatchlings up to near metamorphosis, in which we can actually see the sequence across the metacarpals, metatarsals, and ulna, radius, tibia and fibula. So this is doing it the wrong way round all the way down the limb. This is a totally different mechanism of development, it is ancient, but it doesn't change the adult morphology. Can you explain this?

Duboule: When you say a reversal of structure, you mean the sequence of the cartilage condensation and ossification.

Coates: There is variability distally (in modern salamander limbs) but the chondrification and ossification is not always back to front, relative to the patterns in amniotes and frogs.

Duboule: I don't think this is necessary linked to the patterns of Hox gene expression I showed.

Carroll: It must be a different set of rules.

Duboule: It may be, but as far as I know, in all cases of amphibians that have been looked at, rather comparable patterns of Hox genes were observed. Even in fish pectoral fins, the same patterns are found at an early stage.

Wagner: Well that is not true for the distal Hox gene expression. In urodeles the Hox gene expression pattern is quite different from that in frogs and amniotes.

Also zebrafish does not seem to have a phase III expression as one finds in mouse and chicken.

Duboule: Of course. In the pectoral fin, the growth of the structure is interrupted at some point. We don't know why this is, but we proposed it is because of the fin fold structure, which prevents growth factors signalling from the epithelium to reach the mesoderm. But looking at the early fin bud, everything is there to produce a genuine limb. It is a mechanical problem: at some point the signalling doesn't go through. In salamanders, axolotls and some frogs, comparable distributions of both Hox genes and hedgehog were reported. The morphological result may be different, but at the end there is a polarity.

Hanken: You talked about the digit enhancer as being at a distance from the Hox genes. Has this been demonstrated yet outside the mouse?

Duboule: Yes, it is present in all tetrapods. It is extremely well conserved.

Hanken: I ask because Günter Wagner has argued, in light of the differences that Bob Carroll just mentioned, that the digits in salamanders may have evolved independently from those in other tetrapods. In other words, the tetrapod limb has evolved twice. If so, one might expect that we wouldn't see the exact same mechanisms underlying digit formation in all these groups.

Duboule: This enhancer sequence is large. It is a 40 kb piece of DNA that is incredibly conserved, and is present in all tetrapods. It is also present in *Danio* (zebrafish), coelocanths and tetraodon. The rate of conservation fits well with what we would expect. If the *Danio* enhancer is introduced into mouse though, we don't get expression in digits unlike if we use a tetrapod enhancer.

Hanken: The innovation you claim for tetrapods was not the evolution of the enhancer itself, because it is found in zebrafish.

Duboule: No, in zebrafish we find what we call the global control region (GCR). It is a 40 kb region that contains multiple enhancer sequences. Within this GCR we can see several boxes of conservation. One of them is digit specific. If you take the fish GCR you will find that this box is the least conserved, yet we can see some conservation. If you bring it into the mouse it doesn't work in digits. I would tend to say that this enhancer doesn't work in *Danio*.

Hanken: One of your opening slides showed that if you eliminate HoxA and D clusters you get a runt of a humerus. I noticed, however, that the scapula was fully formed. This is interesting.

Duboule: The piece of that is left, the scapula and mid humeral part, is roughly up to the deltoid crest. This corresponds to the part of the early limb bud that expresses a transcription factor called Meis1. This is a cofactor for Hox function and there is an antagonism between Meis1 and the posterior Hox genes. The interest here is that Meis1 is the gene in vertebrates that is orthologous to homothorax in insects. Homothorax is a gene that is expressed in the insect coxopodite. In insects, the limbs are composed of a sort of trunk extension, the coxa, and of a most distal piece, the telopodite. Insect trunk extension is hedgehog-independent but homothorax-dependent. The early phase of Hox gene activation is hedgehog-independent and is Meis1-dependent. This is why in 2005 we proposed that there is a difference between the morphological and the genetic definitions of a limb (Kmita et al 2005), and the limb starts at the mid-humerus part. I had interesting discussions with morphologists about this! I would argue that this part of our limb is the remnant of the arthropod coxa. It is not a limb, but more a kind of a trunk extension.

Wagner: Is Meis1 working without an association with the 5′ Hox genes?

Duboule: Meis1 is associated with the 3′ Hox genes but its transcription is repressed by the 5′ Hox genes products.

Wagner: But Meis1 is physically interacting with A13, D13 and so on, isn't it?

Duboule: It has been shown by Miguel Torres and Juan-Carlos Izpisua-Belmonte that if Hox group 13 genes are overexpressed, Meis1 is down-regulated. There is a strong expression boundary between these two transcription factors.

Morriss-Kay: In terms of *bricolage*, what is the difference between a fish fin and a tetrapod limb? Is it a small change in the GCR? You said that the fish GCR won't make digits in the mouse, but it is present in the fish.

Duboule: My own view is that the fish GCR lost the capacity to work because there was a morphological transformation of the fin, with the folding of the ectoderm to allow for crest cells to come in and make the exoskeleton. This has a strong adaptive advantage in an aquatic environment. Reducing the endoskeleton and pushing the exoskeleton stops signalling from the ectoderm. This reduces the role of the endoskeleton to the first phase of Hox gene expression and thus reduces the function of hedgehog. It is still there, but it is no longer needed for the endoskeletal compartment

Wilkins: I have a comment on the philosophical considerations you raised rather than the details of the limb. I agree with your account of our idea of transitionism, and I still hold with it. I would add, however, that the best formulation of transitionism I have ever read was in a book that came out 14 years earlier, called *The Biology and Evolution of Language* by Philip Lieberman. He doesn't use the term transitionism, and his focus was morphological rather than molecular-genetic, but the same idea of cumulative small changes leading to qualitatively new properties when critical thresholds are passed is there. With regard to your last comments about *bricolage*, however, I have to say that I think they are not entirely correct because they leave out the population transformation aspect that is essential to evolution. There must be elements of *bricolage* created all the time in all populations through mutations that alter development. The ones that persist and are perpetuated, however, can do this either because they are neutral, and through drift persist, or because they enjoy a selective advantage. I don't think it is just Darwinian storytelling to attribute selective functions to these things. I would suggest that in the particular case you have argued, there probably was some selective advantage in the AP symmetry that may have been created in just the way you claim. But, of course, that would be hard to test.

Wagner: You talked about how the global enhancer and the digit enhancer causes the polarity of the Hox expression in the digit region. How does this mechanism interact with the sonic hedgehog polarity? It gets its identity through hedgehog, so there must be some link between hedgehog, Gli3 and Hox gene activity.

Duboule: It has been shown that the Hox regulates hedgehog likely by direct binding to the hedgehog enhancer that is located a megabase away from the gene. On the other hand, you mentioned the Gli3 effector of the hedgehog signalling pathway. The only genetic mutants we have where this early collinear phase is clearly disrupted is indeed Gli3.

Wagner: How does the digit expression depend on sonic hedgehog?

Duboule: It doesn't depend on sonic hedgehog, but it is modified by it. If sonic hedgehog and Gli3 are removed, you get perfect Hox expression, although it is not polarized, as shown by John Fallon and Rolf Zeller's groups. Hox is repressed by Gli3 and sonic hedgehog de-represses this. But you don't need sonic hedgehog to activate Hox.

Reference

Kmita M, Tarchini B, Zakany J, Logan M, Tabin CJ, Duboule D 2005 Early developmental arrest of mammalian limbs lacking HoxA/HoxD gene function. Nature 435:1113–1116

Affecting tooth morphology and renewal by fine-tuning the signals mediating cell and tissue interactions

Irma Thesleff, Elina Järvinen and Marika Suomalainen

Developmental Biology Research Program, Institute of Biotechnology, POB 56, 00014 University of Helsinki, Finland

Abstract. Interactions between the epithelial and mesenchymal tissue components of developing teeth regulate morphogenesis and cell differentiation, and determine key features of dentitions and individual teeth such as the number, size, shape and formation of dental hard tissues. Tissue interactions are mediated by signal molecules belonging mostly to four conserved families: transforming growth factor (TGF)β, Wnt, fibroblast growth factor (FGF) and Hedgehog. Recent work from our laboratory has demonstrated that tooth morphology and the capacity of the teeth to grow and renew can be affected by tinkering with these signal pathways in transgenic mice. The continuous growth of the mouse incisors, as well as their subdivision into the crown and root domains, is dramatically altered by modulating a network of FGF and two TGFβ signals, bone morphogenetic protein (BMP) and Activin. This network is responsible for the regulation of the maintenance, proliferation and differentiation of epithelial stem cells that are responsible for growth and enamel production. On the other hand, the activation of the Wnt signalling pathway induces continuous renewal of mouse teeth (which normally are not replaced), resembling tooth replacement in other vertebrates. It can be concluded that the different dental characters are quite flexible and that they are regulated by the same conserved signal pathways. These findings support the suggestions that tinkering with the signal pathways is the key mechanism underlying the morphological evolution of teeth as well as other organs.

2007 Tinkering: the microevolution of development. Wiley, Chichester (Novartis Foundation Symposium 284) p 142–157

Epithelial–mesenchymal interactions are mediated by conserved signalling molecules

The communication between cells and tissues is a major mechanism in the regulation of embryonic development, and it has been proposed that the inductive cell and tissue interactions that change the behaviour of the neighbouring tissue constitute the single most important mechanism regulating embryonic development (Gurdon 1987). In the majority of organs, including the teeth, key interactions regulating morphogenesis occur between the epithelial and mesen-chymal tissues. Teeth develop as appendages of the embryonic surface ectoderm, and their early development shares marked morphological as well as molecular similarities with other ectodermal organs such as hair, feathers, scales, nails and many exocrine glands (Pispa & Thesleff 2003).

The central importance of epithelial–mesenchymal interactions for tooth devel-opment has been demonstrated in many classical studies in which the epithelial and mesenchymal tissues of developing teeth were separated and cultured in dif-ferent heterotypic and heterochronic combinations (Kollar & Baird 1970, Ruch 1987, Mina & Kollar 1987, Lumsden 1988). These studies, together with some more recent data, indicate that the tissue interactions take place between the oral epithelium and the underlying neural crest-derived mesenchyme, and that there is a chain of interactive events between the two tissues regulating practically all aspects of tooth development including initiation, morphogenesis, and differentia-tion of the cells forming the dental hard tissues (reviewed in Jernvall & Thesleff 2000, Thesleff & Nieminen 2005).

The elucidation of the molecular basis of the mediation of cell and tissue interac-tions is one of the great breakthroughs in developmental biology research in the last 15 years. The mediators are signal molecules produced by the inducing cells that bind to specific receptors at the surface of the responding cells. Receptor activation initiates the transmission of the signal to the nucleus and results in changes in the expression of target genes. These signal pathways have been conserved throughout animal evolution to a remarkable extent: the same signals regulate the development of all animals and all organs and tissues, and thus the signal pathway genes belong to the common tool-kit of developmental regulatory genes. There are four families of signal molecules which are indispensable for the development of practically all organs, including the teeth. These families are the TGFβ (transforming growth factor β) superfamily, which includes Activin and BMPs (bone morphogenetic pro-teins), the Wnt family, the FGF family (fibroblast growth factors), and the hedgehog family, of which only Shh (sonic hedgehog) is expressed in teeth. Shh together with several signals belonging to the other three families are dynamically expressed during tooth development, and their respective receptors are often seen in the adja-cent tissue, suggesting roles in the mediation of epithelial–mesenchymal interac-tions (Thesleff 2003, *http://bite-it.helsinki.fi*). Many of the suggested roles of signals in the mediation of epithelial–mesenchymal interactions have been confirmed in functional studies (Wang & Thesleff 2005, see below).

Tooth morphogenesis and the diversity of vertebrate dentitions

The central anatomic and genetic features of the morphogenesis of individual teeth are basically similar in all vertebrates and include the formation of the epithelial placode, the budding of the epithelium, the condensation of mesenchymal cells around the bud, and the folding and growth of the epithelium generating the shape of the tooth crown (Fig. 1). This early development of teeth is characterized by the reiterative formation of signalling centres in the dental epithelium that regulate morphogenesis (Jernvall & Thesleff 2000). Such centres form first in the placodes and subsequently in the enamel knots and their functions are carried out by at least a dozen different signal molecules, which are expressed locally in the centres and regulate patterning (Fig. 1). Also the mineralized structures characteristic for teeth (i.e. enamel, dentin and cementum) are formed in different vertebrates by similar specialized cells, the epithelial ameloblasts and the mesenchymal odontoblasts and cementoblasts, respectively.

However, the sizes and shapes of teeth vary greatly, and so do the ways in which dentitions are organized in different animals. Most fish, amphibians and reptiles have a homodont dentition (all teeth have similar shapes), and their teeth are replaced throughout life. The mammalian teeth, on the other hand, are sequentially organized in the jaws into four groups showing characteristic differences in morphology (heterodonty). These teeth are from front to back: incisors, canines, premolars and molars. The mammalian teeth are mostly replaced once (primary, or milk teeth, are replaced by secondary teeth). There are, however, extensive modifications in the dental formulae within mammals, and because the dentition is characteristic for each species, the variations in the patterning, number and

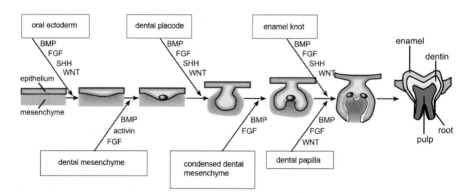

FIG. 1.   Signals belonging to the families of BMP (bone morphogenetic protein), FGF (fibroblast growth factor), Shh (sonic hedgehog), and WNT regulate early tooth development. The signals mediate sequential and reciprocal interactions between the oral epithelium and underlying mesenchymal tissue. Epithelial signalling centres, the dental placode and enamel knots, regulate morphogenesis: the formation of the epithelial tooth bud and its folding morphogenesis resulting in the formation of the crown of the tooth. Epithelial cells differentiate into ameloblasts (enamel forming cells) and mesenchymal cells into odontoblasts (dentin forming cells).

shape of teeth have formed the basis for the analysis of fossil records and understanding mammalian evolution.

Almost all knowledge of the molecular and genetic aspects of tooth development comes from studies on mouse teeth. However, the mouse dentition differs from the dentition of most other animals in several aspects. First, they have only one incisor in each half of the jaw (most mammals have two or three), and the mouse incisors grow continuously. Secondly, although mice have three molars like most other mammals, they completely lack cuspids and premolars. Thirdly, most mammals have two dentitions, but mice have only one dentition, which is not replaced.

Modulation of signal pathways in transgenic mice

Due to the advances in gene technology and the availability of informative mouse models in which the developmental regulatory genes have been modified, we are now beginning to understand in some detail the molecular and genetic basis of tooth morphogenesis. Tooth development is arrested in many knockout mice when the functions of various components of the four essential signal pathways have been deleted. In most cases so far, the deleted molecules have been obligatory for signal transduction, and their deletion has therefore resulted in a complete arrest of tooth development, mostly either prior to placode formation and budding, or before morphogenesis of the bud into cap stage (Fig. 1, Wang & Thesleff 2005). These studies have revealed the necessary functions of many genes and the respective signal pathways, but they usually have not allowed studies on the effects that more subtle tinkering with signalling intensities might have on development.

The very important roles that the fine-tuning of signal pathways have on embryonic morphogenesis have emerged only during recent years. Multiple specific inhibitor molecules have been discovered for each of the four conserved signal pathways, and some of them function in more than one of the pathways. In addition, it has become evident that co-operation of several inhibitors is very common. This presents obvious challenges in the analysis of the inhibitors since simple knockout experiments are often not informative because of functional redundancy. The deletion of a single inhibitor may cause a very mild phenotype that often depends on the genetic background of the mice. Perhaps the most thoroughly studied example of the complexity of signal regulation is the concerted action of multiple TGFβ signals and their inhibitors during dorsal–ventral patterning of the *Xenopus* embryo (Reversade et al 2005). There is also increasing evidence of the important functions of signal modulation in tooth morphogenesis. In particular, fine-tuning of the signalling functions of the dental placodes and enamel knots has dramatic effects on morphogenesis. Tooth number and size, as well as the patterning of the tooth crowns, were shown to be affected in transgenic mice when the levels of signalling were altered by overexpression or lack of the signal modulators Ectodysplasin, Follistatin or Ectodin (Kangas et al 2004, Wang et al 2004a, Kassai et al 2005). Ectodysplasin is a tumour necrosis factor, which stimulates the expression of BMP inhibitors and Shh (Pummila et al 2007), Follistatin is an inhibitor of Activin and BMPs, and Ectodin inhibits BMP and may also modulate Wnt signalling.

Fine-tuning of Wnt signalling affects the capacity of tooth renewal

The capacity of tooth renewal varies between animals but the underlying cellular and molecular mechanisms have remained poorly understood, mainly because there is no tooth replacement in mice (Huysseune & Thesleff 2004). However, we were recently able to produce mouse mutants that renew their teeth continuously. In these mice Wnt signalling was activated in the embryonic ectoderm by conditional expression of a stabilized form of β-catenin ($\beta cat^{ex3K14/+}$, Järvinen et al 2006). The mice died at birth and although the tooth buds were morphologically abnormal, there was no indication of new tooth formation. The remarkable phenotype became evident when the embryonic tooth buds were grown as transplants under the kidney capsules of nude mice. After three weeks of culture, one tooth had generated over 40 new teeth (Fig. 2A). The teeth represented different developmental stages indicating continuous tooth production.

It was shown that the initiation of new teeth in the $\beta cat^{ex3K14/+}$ mice took place sequentially from new enamel knots which were induced in the dental epithelium of the earlier formed teeth. Histological observations of tooth replacement in reptiles that renew their teeth continuously, as well as of human permanent teeth forming from deciduous teeth, have indicated that the successional teeth are initiated from the dental epithelium of their predecessors (Osborn 1971, Fig. 2B). Hence, the process that was activated in the $\beta cat^{ex3K14/+}$ mouse mutants resembles physiological tooth replacement. Apparently, tooth replacement is normally inhibited in mice and in the transgenic mice this inhibition had been relieved. Since the continuous production of teeth in the mutants was caused by activated Wnt signalling in the epithelium, it was concluded that the capacity of tooth replacement depends on the activity of Wnt signalling, and that the lack of secondary tooth formation in normal mice may be associated with an inhibition of the Wnt signal pathway (Järvinen et al 2006).

The flexibility of the tooth replacement process is exemplified by a rare human syndrome, cleidocranial dysplasia (CCD), characterized by multiple supernumerary teeth, which, interestingly, represent a partial third dentition (Jensen & Kreiborg 1990). CCD is caused by heterozygous loss-of-function mutations of the *Runx2* gene (Mundlos et al 1997). Although there is no direct evidence, as yet, of

A

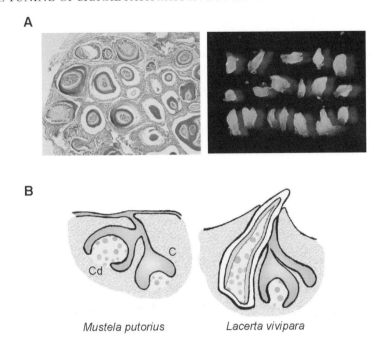

B

Mustela putorius   *Lacerta vivipara*

FIG. 2.   Activation of Wnt signalling in the mouse dental epithelium leads to continuous tooth generation, which resembles tooth renewal in other mammalian species and reptiles. (A) $\beta cat^{ex3K14/+}$ mouse molar tooth grown as an *ex vivo* transplant under the kidney capsule of a nude mouse resulted in a geode-like structure with over 40 individual teeth shown in a histological section and as dissected individual teeth. (B) Successional teeth are initiated from the dental epithelium of their predecessors in mammals (*Mustela putorius*, our unpublished observations) and in reptiles (*Lacerta vivipara*, adapted from Osborn 1971). Cd, primary canine; C permanent canine.

a link between CCD phenotype and Wnt signalling, the observation that Wnt regulates *Runx2* expression in dental mesenchyme (our unpublished results) is in line with this possibility.

Additional support for the role of the Wnt pathway in human tooth replacement comes from the tooth phenotype caused by mutations in *Axin2*, a regulator and direct target of Wnt signalling (Lammi et al 2004). These patients characteristically lack multiple secondary teeth, whereas their deciduous dentition is unaffected suggesting that tooth renewal may be regulated by Wnt signalling (Lammi et al 2004). A role of Wnt signalling in tooth replacement would also be in line with similar functions of Wnts in hair cycling (Gat et al 1998). Because the molecular mechanisms regulating the embryonic morphogenesis of different ectodermal organs are similar (Pispa & Thesleff 2003), it is also conceivable that the mechanisms involved in adult regeneration may be shared between teeth and other organs forming as appendages of the ectoderm. In conclusion, the current evidence indicates that fine-tuning of Wnt signalling plays a key role in tooth renewal, and it can be suggested that the reduced ability of tooth regeneration during vertebrate evolution has involved tinkering with Wnt signalling.

Tinkering with the balance of inhibitory and stimulatory signals affects the growth and enamel patterning of the mouse incisor

The rodent incisors have two specific features that are not shared with most other mammalian teeth: they grow continuously and enamel is deposited asymmetrically so that only the labial (anterior) surface is covered by enamel, whereas the lingual (posterior) surface is covered by dentin and cementum and resembles the roots of molars (Fig. 3B). The continuous growth requires the existence of stem cells capable of generating the differentiated cells producing dental hard tissues throughout the life of the animals. A stem cell niche housing the epithelial stem cells is present in mouse incisors in the cervical loops (Harada et al 1999). The stem cells reside in the stellate reticulum compartment surrounded by a single layer of basal epithelium of the cervical loop (Fig. 3). The stem cells proliferate, invade the basal epithelium, migrate towards the tooth apex, and differentiate into enamel-forming ameloblasts. Interestingly, the morphological features of the epithelial stem cell niche in mouse incisors resemble the two most actively studied adult epithelial stem cell niches, namely the crypt in the intestine and the bulge in hair follicles (Moore & Lemischka 2006). Similar to the intestinal crypt and the hair bulge, the epithelial cell compartment is surrounded by mesenchymal cells, which provide important regulatory signals for the self-renewal and differentiation of epithelial stem cells.

FGF3 and FGF10 were identified as the key mesenchymal signals, which cooperatively stimulate epithelial proliferation in the mouse incisors (Harada et al 1999, 2002, Wang et al 2007). Their effects are modulated by two mesenchymal TGFβ family signals: BMP4 inhibits the expression of *Fgf3*, and Activin counteracts the effect of BMP4 thus stimulating proliferation and the continuous supply of progenitors for ameloblasts (Wang et al 2007, Fig. 4A). The labial–lingual asymmetry in the incisor results from the dramatic size difference in the cervical loops; the labial cervical loop is large and contains abundant stem cells, whereas the thin lingual cervical loop houses only few stem cells. This difference accounts for the labial–lingual difference in growth and enamel formation and is determined by the balance of Activin and BMP. The asymmetric intense expression of Activin labially prevents the inhibitory effect of BMP4 on stem cell proliferation. In addition, the TGFβ inhibitor Follistatin, which is expressed preferentially in the lingual epithelium, prevents the function of Activin in the lingual stem cell niche thus allowing

A

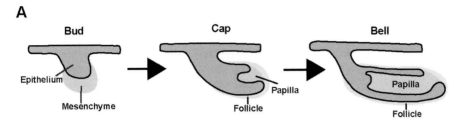

B

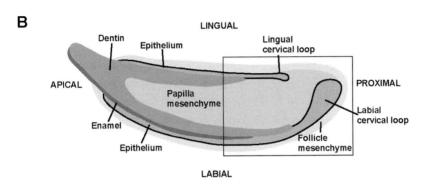

FIG. 3. Development of the asymmetry of the continuously growing mouse incisor. (A) Incisors develop in a morphologically similarly way to molars (Fig. 1) until the labial-lingual asymmetry in growth of the epithelial cervical loops becomes evident during the cap stage. (B) The structure of the incisor as the tooth is erupting. Epithelial ameloblasts have differentiated and formed enamel on the labial but not lingual side. Mesenchymal odontoblasts have produced dentin on both sides.

BMP4 to inhibit *Fgf3* and in this manner reduce stem cell proliferation (Fig. 4A, Wang et al 2007). Interestingly, in the zone where epithelial cells differentiate into ameloblasts, Follistatin inhibits lingually the inductive effect of BMP4 and thereby prevents ameloblast differentiation and enamel formation (Wang et al 2004b).

The series of transgenic mouse phenotypes in Fig. 4B illustrates the dramatic consequences that tinkering with the balance of the stimulatory and inhibitory signals has on the incisor phenotypes. Overexpression of Follistatin (under the K14 promoter) prevents stem cell proliferation (by inhibiting the effect of Activin) and the incisor remains small. Follistatin also inhibits the inductive effect of BMP4 on ameloblast differentiation and therefore no enamel is formed. The deletion of Follistatin function has opposite effects: it increases the size of the stem cell compartment lingually, and stimulates enamel formation on the lingual surface. Inhibition of FGF function in $Fgf3^{-/-}$; $Fgf10^{+/-}$ compound mutants causes reduced growth and severe enamel hypoplasia. Finally, overexpression of the BMP inhibitor Noggin causes hyperproliferation of the stem and progenitor cells in the

A

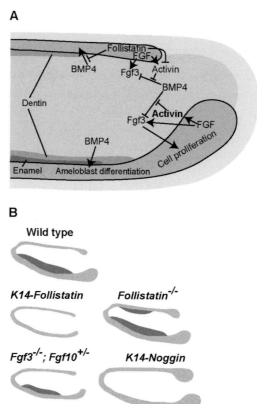

FIG. 4. The network of signal pathways regulating the proliferation of the epithelial stem cells in the mouse incisor, and effects of signal modulation on growth and patterning. (A) Signal network regulating the stem cell niche in the cervical loop (see text for details). Enlargement of boxed region in Fig. 3B. (B) The sizes of cervical loops harboring the stem cells, the growth of the incisor, and the distribution of enamel (dark grey) are dramatically affected as a consequence of altered signal pathways in mutant mice. K14, keratin14 promoter. Noggin is a BMP inhibitor.

cervical loops resulting in remarkable overgrowth of incisors (Fig. 4B). These incisors have no enamel because the BMP-induced ameloblast differentiation is prevented (Plikus et al 2005).

The remarkable variation in the incisor phenotypes of the transgenic mice implicates a key role for the complex signal network in the regulation of the epithelial stem cell niche that determines the growth capacity and the labial–lingual patterning of the incisor. Interestingly, the labial and lingual sides are considered to represent the crown and root domains of rooted teeth, respectively. These observations, therefore, have implications for the regulation of the proximo-distal patterning of rooted teeth, i.e. the determination of the relative heights of the crown and root. The transition from crown to root formation is associated with the depletion of stem cells inside the cervical loop and its timing determines the degree of hypsodonty (Tummers & Thesleff 2003), a very flexible character during the evolution of teeth. High crowned as well as continuously growing teeth are common in many animals, and even primates possess ever growing incisors as seen in the Aye-aye lemurs (Mittermeier et al 1994). It was shown that the continuously growing molars in voles have anatomically similar cervical loops as incisors, and components of the same molecular signal network are in place (Tummers & Thesleff 2003). It is conceivable that fine-tuning of the complex signal network which regulates the maintenance, proliferation and differentiation of the epithelial stem cells has created the evolutionary variation in the crown height and the growth capacity of teeth.

Conclusions

Current evidence indicates that most characteristics of teeth can be manipulated by modulating the four major conserved signal pathways that mediate epithelial–mesenchymal interactions in developing teeth. The numbers of teeth as well as molar cusp patterns can be modified by tinkering with BMP and FGF signalling (Kangas et al 2004, Wang et al 2004a, Kassai et al 2005, Klein et al 2006). The work presented here indicates that the capacity for tooth renewal and replacement is dramatically stimulated by the activation of Wnt pathway in transgenic mice (Järvinen et al 2006). In addition, the continuous growth and asymmetry of the incisor, the formation of enamel, and the subdivision of teeth into crown and root domains can all be significantly affected by modulating a complex network of FGF, BMP and Activin signals that regulates dental epithelial stem cells (Harada et al 2002, Tummers & Thesleff 2003, Yokohama-Tamaki et al 2006, Wang et al 2007). All these findings support the hypothesis that the diversity of tooth types and dental patterns during evolution may have resulted from tinkering with the conserved signal pathways.

References

Gat U, DasGupta R, Degenstein L, Fuchs E 1998 De Novo hair follicle morphogenesis and hair tumors in mice expressing a truncated beta-catenin in skin. Cell 95:605–614
Gurdon JB 1987 Embryonic induction–molecular prospects. Development 99:285–306
Harada H, Kettunen P, Jung H-S, Mustonen T, Wang YA, Thesleff I 1999 Localization of putative stem cells in dental epithelium and their association with Notch and FGF signalling. J Cell Biol 147:105–120
Harada H, Toyono T, Toyoshima K, Ohuchi H 2002 FGF10 maintains stem cell population during mouse incisor development. Connect Tissue Res 43:201–204

Huysseune A, Thesleff I 2004 Continuous tooth replacement: the possible involvement of epithelial stem cells. Bioessays 26:665–671

Järvinen E, Närhi K, Birchmeier W, Taketo MM, Jernvall J, Thesleff I 2006 Continuous tooth generation in mouse is induced by activated Wnt/βcatenin signalling. Proc Natl Acad Sci USA 103:18627–18632

Jensen BL, Kreiborg S 1990 Development of the dentition in cleidocranial dysplasia. J Oral Pathol Med 19:89–93

Jernvall J, Thesleff I 2000 Reiterative signalling and patterning in mammalian tooth morphogenesis. Mech Dev 92:19–29

Kangas AT, Evans AR, Thesleff I, Jernvall J 2004 Nonindependence of mammalian dental characters. Nature 432:211–214

Kassai Y, Munne P, Hotta Y et al 2005 Regulation of mammalian tooth cusp patterning by ectodin. Science 309:2067–2070

Klein OD, Minowada G, Peterkova R et al 2006 Sprouty genes control diastema tooth development via bidirectional antagonism of epithelial-mesenchymal FGF signaling. Dev Cell 11:181–190

Kollar EJ, Baird GR 1970 Tissue interactions in embryonic mouse tooth germs. II. The inductive role of the dental papilla. J Embryol Exp Morphol 24:173–186

Lammi L, Arte S, Somer M et al 2004 Mutations in AXIN2 cause familial tooth agenesis and predispose to colorectal cancer. Am J Hum Genet 74:1043–1050

Lumsden AG 1988 Spatial organization of the epithelium and the role of neural crest cells in the initiation of the mammalian tooth germ. Development 103 Suppl:155–169

Mina M, Kollar EJ 1987 The induction of odontogenesis in non-dental mesenchyme combined with early murine mandibular arch epithelium. Arch Oral Biol 32:123–127

Mittermeier RA, Tattersall I, Konstant WR, Meyers DM, Mast RB 1994 Lemurs of Madagascar. Conservation Intl Washington, DC, USA

Moore KA, Lemischka IR 2006 Stem cells and their niches. Science 311:1880–1885

Mundlos S, Otto F, Mundlos C et al 1997 Mutations involving the transcription factor CBFA1 cause cleidocranial dysplasia. Cell 89:677–680

Osborn JW 1971 The ontogeny of tooth succession in *Lacertia vivipara* Jacquin (1787). Proc R Soc Lond B Biol Sci 179:261–289

Pispa J, Thesleff I 2003 Mechanisms of ectodermal organogenesis. Dev Biol 262:195–205

Plikus MV, Zeichner-David M, Mayer JA et al 2005 Morphoregulation of teeth: modulating the number, size, shape and differentiation by tuning Bmp activity. Evol Dev 7:440–457

Pummila M, Fliniaux I, Jaatinen R et al 2007 Ectodysplasin has a dual role in ectodermal organogenesis: inhibition of BMP activity and induction of Shh expression. Development 134:117–125

Reversade B, De Robertis EM 2005 Regulation of ADMP and BMP2/4/7 at opposite embryonic poles generates a self-regulating morphogenetic field. Cell 123:1147–1160

Ruch JV 1987 Determinisms of odontogenesis. Revis Biol Celular 14:1–99

Thesleff I 2003 Epithelial-mesenchymal signalling regulating tooth morphogenesis. J Cell Sci 116:1647–1648

Thesleff I, Nieminen P 2005 Tooth induction. In: 'Nature Encyclopedia of Life Sciences'. Also online http://els.wiley.com

Tummers M, Thesleff I 2003 Root or crown: a developmental choice orchestrated by the differential regulation of the epithelial stem cell niche in the tooth of two rodent species. Development 130:1049–1057

Wang X-P, Thesleff I 2005 Tooth development. In: Unsicker K, Krieglstein K (eds) Cell signalling and growth factors in development. Wiley-Vch Verlag GmbH&Co KGaA, Weinheim, p 719–754

Wang X-P Suomalainen M, Jorgez CJ et al 2004a Modulation of activin/bone morphogenetic protein signalling by follistatin is required for the morphogenesis of mouse molar teeth. Dev Dyn 231:98–108

Wang X-P, Suomalainen M, Jorgez CJ, Matzuk MM, Werner S, Thesleff I 2004b Follistatin regulates enamel patterning in mouse incisors by asymmetrically inhibiting BMP signalling and ameloblast differentiation. Dev Cell 7:719–730

Wang X-P, Suomalainen M, Felszeghy S et al 2007 An integrated gene regulatory network controls epithelial stem cell proliferation in teeth. PLoS Biol, in press

Yokohama-Tamaki T, Ohshima H, Fujiwara N et al 2006 Cessation of Fgf10 signaling, resulting in a defective dental epithelial stem cell compartment, leads to the transition from crown to root formation. Development 133:1359–1366

DISCUSSION

Lieberman: Is there evidence for the same pathways in other species?

Thesleff: The molecules regulating tooth development have not been analysed in too many species, but the evidence from those that have been studied suggests that, in general, the same pathways are involved. The Wnt pathway has been associated with tooth renewal in humans: the excessive formation of rudimentary teeth in tumour-like outgrowths in human jaws is seen is some cases of the APC syndrome which is caused by mutations in the Wnt pathway (Fader et al 1962), and mutations of Axin2 another Wnt signal mediator result in inhibition of tooth renewal (Lammi et al 2002). Also in chicken, the activation of Wnt pathway induces tooth like structures (see below in discussion). We have analysed the pathways associated with continuous tooth growth in voles. Interestingly, the same pathways as in mouse incisors were in place not only in the vole incisors but also in the continuously growing molars (Tummers & Thesleff 2003).

Oxnard: In earlier days I was in the University of Chicago department with Ed Kollar. I felt that his original work was excellent. I know, however, that he was incredibly depressed that it was not well received nor understood. I'm delighted his work has been mentioned very positively here.

I have a second plea: could work of this type be done on a species that has lots of different kinds of teeth? Two types of teeth only as in rodents is very limiting. Would it also be important to try this kind of work in another anatomical location? The kidney, for example, is well known for producing ectopic hard materials (e.g. bone, teeth). There just might be something in that environment that is also involved.

Thesleff: Hard tissue production is a different issue from morphogenesis. Perhaps the other calcification processes elsewhere may not be related to tooth production.

Hallgrimsson: Teeth can develop in kidneys.

Duboule: That was one of the original observations of Roy Stevens working with teratocarcinomas: he saw the spontaneous occurrence of teeth in testes tumours.

Thesleff: Yes, it is true that teeth are seen also in ovarian teratomas in so called dermoid cysts together with other tissues such as hair and bone, and these are derived from germ cells.

Duboule: There are quite a number of mutant stocks in mice where the face is a little bit skewed for one reason or another, leading to a wrong alignment of incisors; they do not meet. These mice die rapidly because they can't eat and the incisors grow and penetrate the palate. How do you think this could evolve? You need a precise contact point between the teeth if you want the mouse to survive. How can you think of a process that would give this extraordinary meeting point?

Thesleff: This sort of phenotype is usually seen with asymmetric development of the mandible or other cranial structures. During normal development the position of the incisors is strictly controlled.

Duboule: Are there rodents which don't have continual incisor growth?

Thesleff: It is found in all rodents, but it is rare in other animals. There is a species of lemurs that has this, so it isn't limited to rodents. In addition there are continuously growing molars in many different animals including some rodent species. What I suggest here is that the signalling network that I described in the continuously growing mouse incisors may be modulated in different animals and in different types of teeth, and that the balance on the signalling system determines whether the stem cells required for continuous growth are maintained or not.

Jernvall: Continuously growing incisors is a trait that has evolved in at least half a dozen situations in different mammalian lineages. It almost always involves a higher number of incisors that is then reduced, eventually leading to the evolution of continuously growing incisors. It is very much the overall shape of the skull that is quite similar in these mammals. This is something that can be done by tinkering. It doesn't have to happen suddenly because intermediates can be found. This is the same with cheek teeth: the teeth become taller and taller, and after a point they form no roots.

Duboule: So the general view is not that at some point there were non-continuously growing incisors that met at a point, and then once these teeth were allowed continuous growth was allowed.

Jernvall: Yes, continuous growth evolved after the evolution of occlusion.

Weiss: The occlusion isn't developed based on physical contact. The growing upper and lower teeth are not in contact and so don't mould to each other. So, somehow, as the morphology of molars evolves, they have to evolve occlusability, because they don't meet until after they erupt, which is after they have already developed.

Cheverud: Then, when they do meet, they move. There are compensatory mechanisms involved too. How they arise in development would be the stabilizing selection, with the two parts needing to work together. For opposing teeth, they have to have close enough contact to adjust. This would be a strong selection for pleiotropy between the two.

Weiss: Someone asked whether these genes have been seen in other species. I know that in bird jaws, more-or-less similar patterns have been found. David Stock has worked in fish and has found similarities in both jaw teeth and pharyngeal teeth, but they are not identical. So you could say that the same kit of genes is being expressed in slightly different areas and at slightly different times in different lineages. The general pattern is old, the details changed.

Donoghue: The similarities are general at the level of epithelial appendages, rather then being teeth specific. There is a generic tool kit for patterning epithelial appendages.

Lieberman: Irma Thesleff mentioned the importance of inductive interactions. Teeth have to continually reposition themselves during growth for proper occlusion. This is one of those systems that remain incredibly labile and highly epigenetic throughout development. As skulls change shape those teeth have to be able to move to be in precise occlusion or there are serious fitness consequences for the animal. It is a beautiful example of how evolution acts on developmental processes late in ontogeny.

R Raff: Isn't there a general class of problems here that doesn't relate specifically to teeth interacting, but includes interactions beyond immediate tissue interactions? For example, why are my two arms the same length?

Lieberman: Exactly. There is a difference in the degree of asymmetry between flying animals versus swimming animals.

Weiss: There is an even more profound issue, which is that teeth in animals ranging from moles to elephants are of similar types and numbers, but have orders of magnitude differences in size. The mathematical modelling of this has suggested that this has to happen by inductive patterns at an early stage in the embryo. Otherwise the kinds of signalling and patterning mechanisms would go chaotic if you waited until these embryos are bigger.

Coates: David Stock and other groups researching the dentitions of teleost fishes are also beginning to look at the subtle changes in the patterning mechanisms in tooth replacement. The initial tooth patterning systems are not the same as those for subsequent replacement teeth. In sharks there is an anticipation of jaw growth: in larval chondrichthyans, functional teeth in the tiny jaws are mere shards, but the teeth further down the tooth families are much larger anticipating the growth of the jaw.

Hall: There is a conundrum of having individual teeth co-ordinated, but if you have to have the whole dentition co-ordinated this is even more complex. Irma, in your organ culture work in which you were getting waves of teeth that you interpreted as potentially successive waves, was the morphology of each of those waves the same? It would be nice if they were getting more complex as subsequent waves arose.

Thesleff: No, they weren't. I don't think this just mimics the replacement of sets of teeth, but also the process where the teeth in a tooth family are formed successionally.

Hall: Have you looked at ectopic Wnt expression in the diastema region? There are rudimentary tooth buds in the diastema, at least in voles if not in mice.

Thesleff: There was a wonderful recent paper by John Fallon's group (Harris et al 2006) where they showed that by overexpressing Wnt in chicken epithelium they could induce multiple ectopic buds in the oral epithelium resembling early tooth formation. We didn't look at the initiation stage, because we didn't see any phenotypic changes from overexpressing Wnt throughout the ectoderm in mice during tooth initiation. The placodes of incisors and molars looked very normal. The first differences were seen at the bud stage where the contour of the bud became irregular and it started to produce more buds. We didn't see this in the diastema region.

Weiss: These tiny buds that form anterior to the first molar have been argued by some people to be vestigial diastema teeth. A few recent papers have used transgenic alterations in mice to keep these vestiges growing, to form mature teeth that have been interpreted as diastemal teeth (e.g. Peterkova et al 2005, Klein et al 2006, Kassai et al 2005). Of course, when you start disrupting patterning it is hard to decide whether it really is in the diastema region.

Weiss: There is another issue which is important, relating to Brian's comment that these have to develop as a continual field. This isn't really true in the upper dentition. The medial nasal process contains the upper incisors, and the other upper teeth are in the maxillary process of the first mandibular arch. This implies either independent instruction, or that the pre-patterning is laid down very early on, but separately for each of these tissues. There are lots of interesting instructional memory issues involved.

Oxnard: Does anyone know anything about the ontogenetic loss of teeth? If one looks at certain genetic skull anomalies (or even some environmental skull variations) in humans one can find normal first incisors and canines, and then a tiny second incisor. One can have individual teeth from the deciduous dentition sitting amidst permanent teeth even in quite mature adults. What is going on in those sorts of situations?

Thesleff: There is a clear association between the reduction of tooth size and the disappearance of teeth. The maxillary lateral incisor is one of the most commonly missing teeth in humans, and the conical-shaped small incisor is seen in the same families as congenitally missing teeth. Often the same individual may lack the lateral incisor on one side and have a small conical incisor on the other side. Hypodontia, i.e. the lack of teeth, is a typical autosomally dominantly inherited condition and there is always remarkable phenotypic variation between patients in the same family in the number and identity of the missing teeth.

Weiss: But it is usually in the last end of a series. The last molar is usually the most variable and the last one to go. It is like there is a developmental field that runs out of developmental 'energy'.

Oxnard: In some animals there can be as many as six incisors. It is not the end of the series if there are four or five. The same applies for premolars.

Thesleff: It is all about the size of the rudiment as it forms. In molar development the first molar develops from the placode, and the second doesn't have its own placode, but forms from the first molar, and the third molar forms successionally from the second. This is why the third molar is commonly missing.

Hall: So it is the size of the signalling system, a bit like the wing spots in butterflies.

References

Fader M, Kline SN, Spatz SS, Zubrow HJ 1962 Gardner's syndrome (intestina polyposis, osteomas, senaceaous cysts) and a new dental discovery. Oral Surg Oral Med Oral Pathol 15:153–172

Harris MP, Hasso SM, Ferguson MWJ, Fallon JF 2006 Development of archosaurian first-generation teeth in a chicken mutant. Curr Biol 16:371–377

Kassai Y, Munne P, Hotta Y et al 2005 Regulation of mammalian tooth cusp patterning by ectodin. Science 309:2067–2070

Klein O, Minowada G, Peterkova R et al 2006 Sprouty genes control diastema tooth development via bidirectional antagonism of epithelial-mesenchymal FGF signaling. Dev Cell 11:181–190

Lammi L, Arte S, Somer M et al 2004 Mutations in AXIN2 cause familial tooth agenesis and predispose to colorectal cancer. Am J Hum Genet 74:1043–1050

Peterkova R, Lesot H, Viriot L, Peterka M 2005 The supernumerary cheek tooth in tabby/EDA mice—a reminiscence of the premolar in mouse ancestors. Arch Oral Biol 50: 219–225

Tummers M, Thesleff I 2003 Root or crown: a developmental choice orchestrated by the differential regulation of the epithelial stem cell niche in the tooth of two rodent species. Development 130:1049–1057

GENERAL DISCUSSION II

R Raff: What I have noticed during this meeting is that we seem to have trouble deciding what tinkering is. I want to present some issues of organization which we could think about. The first is philosophical. What is the agent? Jacob identifies this agent as natural selection, and he says a very interesting thing in his paper: there is nothing in any human activity that is like natural selection, and the closest thing he can come to is the tinkerer. This is a penetrating statement. The other interesting one is the perversion by modern creationism, which is intelligent design, where they would be quite happy with tinkering. Everything was created at the beginning: all the structures of the cell and the components, and it has been played out since then by the choice of different parts.

The second issue is mechanistic. There we have issues such as co-option, but also, as identified by Jacob, the issue of constraints and contingency—not all space is filled. Then there is a third issue, that of scale. This has emerged in some of our discussions, where some have maintained that tinkering is just small scale changes. I would suggest that tinkering happens at all scales and there is probably some frequency distribution where we go from small to large effects. It may be that there is some scaling involved simply by probability: if you want to grow a new pair of limbs at one time it is a tinkering event, but it is an unlikely one to succeed. It will depend on contingency and various other factors as well. The last of these would be level. I haven't seen much conflict on this: it seems to go from genes to morphology. Certainly, people have identified tinkering-like events at these sorts of levels of organization.

Wagner: Perhaps another dimension we can use to clarify our thinking is as follows. Tinkering can mean a process or a concept. Here you are thinking of a concept. If I am trying to find out what a concept means, the best starting point is to ask to what scientific problem does the concept answer? What compels us to adopt a concept like tinkering? And can we live without it? If we give it up, what would we lose? A concept is a tool, and a tool is defined by what we can do with it.

Lieberman: Another philosophical distinction would be thinking of tinkering as a process rather than a pattern. I think this is what Jacob meant. A pattern is observed without knowledge of the process. In many of the papers presented so far we have been observing differences in developmental processes between distantly related organisms. One interesting question is whether the differences we observe are the real stuff of evolution. Are the differences between closely related organisms fundamentally different from the ones we see among distantly related organisms? Are we looking at the end result of many tinkering changes, or are we looking at changes in scale that are stochastic?

Brakefield: I'd like to bring this back to the tooth story. To me, a fascinating issue is that now you understand in some detail the signalling pathways involved in making a tooth in the mouse. Are you also interested in taking natural populations or selected lines based on large outcrossed populations that have different phenotypes, and asking which of these pathways and genes account for the differences in phenotype at that level?

Jernvall: That is the next step, yes.

Brakefield: What do you predict? What sort of targets within those networks do you think you will find that contribute to those more subtle differences within the species?

Jernvall: We don't yet know whether the identity of the target matters because we can reproduce similar phenotypic effects in mouse tooth using different signals. Is it always going to be the same gene, or is it more the logic of the network?

Brakefield: Once you have this type of answer you can begin to ask whether tinkering at that level really does have the same sort of mechanistic basis as the differences across more widely different species.

Cheverud: The gene effect seen today is not necessarily the gene effect that was there in the beginning. With regard to scale, I think most of the tinkering is done at a small scale. Many of the large scale differences we see might have originated as small differences but which since have been burdened by other features and therefore now appear to have a large effect.

Hall: Is tinkering something that was very common early in the origin of life and has now dropped off? Or has it been going on all through evolutionary time? Are we talking about the same thing when we refer to tinkering?

Cheverud: Evolution always has to happen in populations, and this will always put it at the small timescale. What appear to be bigger differences are the accumulation of macroevolutionary processes that are not changes in means of traits in populations over time, but these effects all have to arise at the level of sister species at some point. What appears big now wasn't necessarily big then.

Bell: Jim, don't you think that Alan Orr (1998) has argued the opposite?

Cheverud: I think he has.

Wagner: Everything Alan is talking about is occurring on a microevolutionary scale.

Bell: Orr's (1998) theory applies beautifully for the major morphological changes we see in the threespine stickleback.

Wagner: I don't think the logic of the model applies.

Wilkins: Can you remind us exactly what Alan Orr proposed?

Wagner: He says that if you have an adaptive process, early on you select genes with larger effects on fitness, and then the next substitutions have smaller effects following a geometric distribution function.

Wilkins: Those large differences in fitness don't necessarily correspond to large morphological changes, however.

Bell: It depends on how fitness maps onto morphology.

R Raff: Let's return to what Jacob meant. He was talking about tinkering with what selection is doing. Are we introducing another word that means selection?

Lieberman: I would say that Jacob is not saying it is selection. He says that selection is an outcome of tinkering, or acts upon tinkering. Selection doesn't produce tinkering.

Brakefield: Tinkering provides the variation which natural selection acts on.

Bard: If you tinker with the environment, you tinker with the selection pressure. It can work both ways.

Duboule: We shouldn't forget the historical perspective. At the time François Jacob wrote his paper, we wouldn't have had any idea that genes could be used in different places during development. The first accounts of such a process came in the late 1980s, in reports saying that in development, important genes would be used in different places, at different times, to do different things.

Lieberman: Let's not forget King & Wilson (1975).

Duboule: Most of us had the chance to witness this period. Before 1987, we would have thought that elephants had trunks because they had trunk genes. If we don't keep this in mind we aren't going to approach 'tinkering' using the meaning Jacob had in mind.

Hall: A lot of Jacob's motivation was that humans and chimps had so many genes in common, so where could the differences have come from?

Duboule: At the time he wrote his paper there were no cloned human genes. The first human cloned gene came in the late 1970s.

Hall: But he knew the King & Wilson paper about the percentage of the genome that was in common.

Weiss: Globin genes were known to be duplicated much earlier than that. The protein sequence was known. The idea of gene families is earlier.

Duboule: This was by global hybridization.

Weiss: I would like to raise the controversial name of Bateson, in the context of a gene for everything. He was observing that meristic characteristics, built of repeated similar units like digits or teeth, were made by a common process rather than individual determinitive entities. Even after Mendel's work was rediscovered and Bateson became a Mendelian, he still argued that many repetitive traits were due to a common process rather than individual particulate elements.

Budd: Thinking about what Adam, Rudi and Denis have said, I am beginning to get some ideas about tinkering that I don't entirely approve of. That is, this idea that development is presenting natural selection a done deal. This is betrayed in language like 'genetic toolkit': the idea that there are modules presented for selection to work with, and it has to make do with that sort of stuff. Is this really the case, or can we think of natural selection penetrating much deeper? That is, could natural selection be going on right at the beginning of this toolkit?

Wilkins: I've never used the term 'genetic toolkit'; I think it is too simple, lacks an evolutionary perspective and is, therefore, ultimately misleading.

Bard: Selection does act within the cell, but the normal negative response would be apoptosis. The trouble is that a positive result would probably lead to cancer!

Budd: In terms of regulatory development, what is providing the adaptive landscape?

Lieberman: Any system where there is competition among units can lead to a kind of natural selection.

Wagner: He is asking where fitness differences are coming from. Some come from the need for functional coherence.

Budd: Rudi talked about the relative roles of downstream genes versus genes acting higher up in pathways. Presumably there is some co-option going between these two.

R Raff: One of the things Jacob identified that was interesting about selection was that as well as co-option, there were also constraints present. Among these are historical constraints or contingencies. If you look at evolution taking place, selection can only work on whatever phenotypic variation is there. The story has been somehow that selection is independent and working with random variation. I think these other concepts suggest that variation has a random element in terms of mutations arising from the genome. In fact, the very structure of the organism provides contingency and constraints. If you go back to tooth buds, you can imagine a tooth bud evolving where Wnt and FGF switch roles in terms of the genes they govern, and the result would be the same in mandible formation. These two signalling systems just happen to have different roles. It is an accident which signalling system got used for which final outcome.

Lieberman: I would add one more category here, which Jacob spends a lot of time discussing in his essay: the consequences of the system. Jacob starts his essay with hopeless monsters where the consequence of this kind of co-option is highly integrated systems with buffering and canalization in addition to the complexity we see. This turns out to be useful for organisms and allows them to survive the slings and arrows of outrageous misfortune during development. He wouldn't have used the term 'emergent property', but there are many aspects of development which we observe that are emergent properties of tinkering.

Wilkins: There are all sorts of potential inherent in organisms because of the way evolution has selected for certain kinds of modules to do certain things. This goes back to the argument I made yesterday: different mutations in different modules can affect the operation of other modules because of the conditional linkages between them. A lot of the order we see, even in mutants where there is some sort of phenotypic 'sense' to what you get, is because there has been some prior evolutionary honing of functional modules by natural selection. New mutations can either suppress expression of modules or evoke the expression of others that were not previously part of that developmental process. I'd like to return to the philosophical point Günter Wagner raised, however. To evaluate the worth of the idea of tinkering, we have to ask what do we lose if we give up this idea? If we do, the only alternative is some sort of engineering. It is either natural selection as the engineer (Jacob says, 'The action of natural selection has often been compared to that of an engineer. This, however, doesn't seem to be a suitable comparison'), or you take God or some supreme being as your engineer. Clearly, however, natural selection doesn't act as an engineer. Ultimately, the significance of what we are talking about here is evidence against the creationist intelligent design ideas that are flying around today.

Wagner: Then it becomes just a label for Darwinian evolution.

Wilkins: It is a bit broader than that; it embodies a philosophical position.

Wagner: So it is not a scientific concept?

Wilkins: I think it is more of a metaphysical concept than a scientific one, but it is useful.

Hall: Jacob tied it explicitly to mechanisms.

Wilkins: I'm talking about tinkering as a general process. Then you have to ask how natural selection uses tinkering, or how tinkering comes in, in specific ways. Those enquiries then bring us to mechanisms.

Hall: Jacob would say it is by changes in regulation of gene action, not by changing the structures of genes.

Olsson: I'm thinking about what the limits of tinkering are. It seems that some people would consider things like orthogenesis or saltationist processes as tinkering.

Hanken: It might help to recall recent debates in evolutionary biology, especially those that prevailed around the time Jacob wrote about tinkering. In the 1970s there was widespread discussion of the concept of 'pan-selectionism' (e.g. Wills 1973). The debate focused on how strong natural selection is and how efficient it is at shaping organisms. I see tinkering as similar to the Gould and Lewontin side of things, which challenges the validity of pan-selectionism. Inherent to tinkering is the idea of limits; these limits are shaped by history and the fact that natural selection is limited by the materials it has to work with.

Hall: I think tinkering was proposed by Jacob for the same reason epigenetics was proposed by Waddington. The emphasis was on genes doing everything, and he wanted to put the emphasis on another level of control. Do we need this term any longer now that we all understand gene structure better?

Wagner: The other red herring was the selfish gene. This was the immediate context against which tinkering was proposed.

References

King MC, Wilson AC 1975 Evolution at two levels in Humans and Chimpanzees. Science 188:107–116

Orr HA 1998 The population genetics of adaptation: the distribution of factors fixed during adaptive evolution. Evolution 52:935–949

Wills C 1973 In defense of naive pan-selectionism. Am Nat 107:23–34

Evolution of covariance in the mammalian skull

Benedikt Hallgrimsson, Daniel E. Lieberman*, Nathan M. Young†, Trish Parsons and Steven Wat

*Department of Cell Biology, and Bone and Joint Institute, University of Calgary, 3330 Hospital Drive NW, Calgary, Alverta T2N 4N1, Canada, *Department of Anthropology, Harvard University, 11 Divinity Avenue, Cambridge, MA 02138, USA, †Department of Surgery, Stanford University, Stanford, CA 94305–5148, USA*

Abstract. The skull is a developmentally complex and highly integrated structure. Integration, which is manifested as covariance among structures, enables the skull and associated soft tissues to maintain function both across ontogeny within individuals and across the ranges of size and shape variation among individuals. Integration also contributes to evolvability by structuring the phenotypic expression of genetic variation. We argue that the pattern of covariation seen in complex phenotypes such as the skull results from the overlaying of variation introduced by developmental and environmental factors at different stages of development. Much like a palimpsest, the covariation structure of an adult skull represents the summed imprint of a succession of effects, each of which leaves a distinctive covariation signal determined by the specific set of developmental interactions involved. Covariance evolves either by altering the variance of one of these sequential effects or through the introduction of a novel covariance producing effect. Either way is consistent with the notion that evolutionary change occurs through tinkering. We illustrate these principles through analyses of how genetic perturbations acting at different developmental stages (embryonic, fetal, and postnatal) influence the covariance structure of adult mouse skulls. As predicted by the model, the results illustrate the intimate relationship between the modulation of variance and the expression of covariance. The results also demonstrate that covariance patterns have a complex relationship to the underlying developmental architecture, thus highlighting problems with making inferences about developmental relationships (e.g. modularity) based on covariation.

2007 Tinkering: the microevolution of development. Wiley, Chichester (Novartis Foundation Symposium 284) p 164–190

I mean by this expression (correlated variation) that the whole organization is so tied together during its growth and development, that when slight variations in any one part occur, and are accumulated through natural selection, other parts become modified. This is a very important subject, most imperfectly understood, and no doubt wholly different classes of facts may be here easily confounded together.
Charles Darwin, The Origin of Species (6th edition, 1872), p 184

Integration is a ubiquitous feature of complex organisms (Olson & Miller 1958). The concept of integration captures the idea that the parts of an organism in a population do not vary independently and that the degrees of non-independence vary among sets of parts. Integration is reflected in the predictable shape transformations that take place during growth, the existence of predictable size-shape relationships among adult structures, and the maintenance of appropriate size relationships among structures that function together (e.g. the upper and lower jaw or limbs perform across a range of overall body sizes). As the above quote and the fairly substantial discussion of correlated variation in *The Origin of Species* illustrates, Darwin clearly appreciated this phenomenon and understood its evolutionary implications. Quantitative analysis of integration dates from the beginning of the modern synthesis, with the debate centring on the existence of developmental or genetic factors affecting overall size versus the size of individual parts (Castle 1929, Wright 1932). Olson and Miller's (1958) systematic treatment defined the modern study of morphological integration and developed the thesis that correlations between structures reflect shared developmental origins or function. Picking up from Olson and Miler, Cheverud (1982, 1988, 1995, 1996) placed the study of integration in a quantitative genetics framework. In a series of papers Cheverud developed his general thesis that through selection acting on the genetic and developmental determinants of pleiotropy, functional integration leads to genetic integration which, in turn, leads to developmental integration.

Although not explicitly discussed in Jacob (1977), integration is a predicted outcome of tinkering in that it leads to modularity. As formulated by Jacob (1977), natural selection essentially takes advantage of existing variation to modify developmental pathways. Tinkering helps account not only for why evolved things tend not to be hopeless monsters, but also for the widespread existence of pleiotropy, linkage, and epigenesis—all of which lead to integration. A common manifestation of such integration is modularity. The concept of modularity in organismal structure and development (Wagner 1995, 1996) has provided a theoretical framework that has since guided the study of integration (Hallgrimsson et al 2002, Klingenberg et al 2003, Klingenberg & Zaklan 2000, Magwene 2001, Young & Hallgrimsson 2005). As Wagner has pointed out, this idea has three parallel origins: as semi-autonomous parts of an organisms (Raff 1996); as distinct components of genetic networks (Gerhart & Kirschner 1997); and as parts of organisms composed of traits that covary because of genetic pleiotropy and developmental interactions (Wagner 1995, 1996). It is this form of modularity, that Wagner and Mezey (2003) term 'variational modularity' that provides the current conceptual framework for understanding integration in organismal structure.

Modularity is a central concept for evolutionary developmental biology and it provides a clear conceptual framework for understanding integration and its relationship to tinkering. Covariation structure is essential in complex organisms because of the need to maintain appropriate size and shape relationships among structures across ontogeny and ranges of phenotypic variation. Covariation structure emerges when genetic or environmental sources of variation affect some structures and not others. If all structures are equally affected, then there is integration, but it is not structured. This might happen for instance if size (and isometric scaling) were the only determinant of covariation. On the other hand, if structures are affected randomly, there is no integration. So, covariation structure reflects the organization of organisms into sets of structures or traits that share developmental or genetic influences. In other words, covariation structure is determined by the (variational) modularity of the organism. Modularity also links integration to tinkering and evolvability (Schlosser & Wagner 2003). As noted above, the partitioning of organisms into variational modules reduces the probability that a mutation which enhances function for one trait has deleterious effects on others.

Despite this solid foundation of theory and much empirical work, little is known about how the modular organization of development actually generates covariation structure in complex phenotypes. In this paper, we present a developmental model for understanding covariation structure in the context of tinkering. We argue that covariation is produced by variation in developmental processes that affect aspects of the phenotype unequally. For complex structures, there can be many such processes, so the eventual covariation structure produced represents the sequential (or time-overlapping) overlay of many such developmental processes. Changes in covariation structure are produced in two ways. Most commonly, an alteration in the variance of one of these covariation-generating processes will produce an altered covariation structure and a change in overall integration. Variation in these processes is essential since without variation there is no covariation and the amount of variation expressed in particular processes determines covariation structure.

Focusing on the covariation structure of the mouse skull, we make the following argument:

1. Integration in the skull (and many other complex structures) is produced by a limited, tractable number of key developmental processes.
2. Covariance structure is the summed effect produced by the variances of these covariance-generating processes.
3. Changes in variance in covariance-generating developmental processes can produce radically different covariance structures.
4. Changes in variance can also increase or decrease the overall magnitude of integration.

This model for the developmental origins of covariance structure makes explicit the relationship between the modulation of variance by development and the phenotypic expression of covariance. In other words, it reveals how canalization and integration interact although changes in the variance of developmental processes need not involve canalization. It also shows that patterns of phenotypic integration have a complex and sometimes indecipherable relationship to the underlying variational modularity that generated them. This does not mean that patterns of phenotypic integration cannot be used to study modularity, but rather that additional information, such as that provided by known developmental perturbations, may be needed to test hypotheses about the developmental determinants of modularity. Most importantly, this model clarifies how covariation structure evolves by alterations to developmental determinants of phenotypic variation. In so doing, our model shows the central importance of key developmental processes in structuring the genotype–phenotype map.

The palimpsest model of integration of the skull

To generate covariance among cranial structures, developmental processes must either affect the entire skull or some portions and not others. Taking a broad view, we can make several hypotheses about the main processes that fulfil these criteria during cranial development (Plate 1). The list below is not meant to be exhaustive, but it is meant to capture the most important covariance-generating processes in the skull.

A. Neural crest migration

Much of the mesenchyme of the head is derived from neural crest that originates at the tips of the neural folds and migrates ventrally alongside the neural tube towards the face and anterior basicranium (Hall & Horstadius 1988). The remainder is derived from somites or head somitomeres. The bones of the face, the anterior cranial base and the anterior cranial vault (frontal bone) are formed from neural crest while the posterior cranial base and posterior cranial vault are derived from mesoderm (Jiang et al 2002, Noden & Trainor 2005). Genetic factors that influence only the neural crest or only the mesoderm-derived mesenchyme and its subsequent proliferation and differentiation would cause covariation among the structures derived from each of these origins.

B. Neural crest patterning and proliferation

Within the neural crest-derived mesenchyme, regions corresponding to the branchial arches as well as subregions within the branchial arches are patterned by the combinatorial expression of members of the *Dlx* homeobox gene family (Depew et al 2002a, Kraus & Lufkin 2006). Subsequently, the fates of these regions vary as subtly different, albeit overlapping, series of developmental genetic events play out on the regionally patterned mesenchyme (Depew et al 2002b). Differences in the local regulation of cell proliferation within the facial mesenchyme enabled by this patterning are key to successful formation of the face and also for producing species-specific facial morphology (Brugmann et al 2006, Helms et al 2005). Structures derived from these regions or subregions thus have common genetic or epigenetic influences that are either not shared or act differently on other regions. Variation in these common influences will cause covariation among structures derived from specific regions of the facial mesenchyme.

C. *The fusion of the facial prominences*
The face forms from prominences that grow outwards around the oropharyngeal membrane. There are initially five prominences—the midline frontonasal process and the paired maxillary and mandibular processes (Plate 2). The frontonasal prominence splits into a medial and lateral nasal prominence on either side of the nasal pits. The formation of the primary palate involves fusion first between the medial and lateral nasal prominences and then between the lateral nasal prominence and the maxillary prominence (Jiang et al 2006). In mice, this takes place between GD 9.5 and GD 12 (Wang & Diewert 1992). This process involves the co-ordination of growth by mesenchymal proliferation within these prominences and their fusion via the growth of filopodial interdigitations at the epithelial seam and subsequent apoptosis or epithelial–mesenchymal transformation of the seam itself (Jiang et al 2006). The fusion of the facial prominences produces a physical contact between the previously separate outgrowths and changes the subsequent dynamic of facial growth and morphogenesis. Variation in the timing, relative degree of outgrowth of the prominences and their subsequent degree of physical contact could thus influence the growth of the face. Genetic variations that influence the fusion of the face could thus contribute to covariation among facial structures.

D. *Mesenchymal condensation and differentiation*
The formation of mesenchymal condensations is a crucial first step in the development of skeletal structures (Hall 2005). Condensations form initially through increased local proliferation of mesenchymal cells, migration towards a central area, the formation of filopodial connections between cells, and the expression of cell adhesion molecules such as N-Cam (Hall 2005). This process if followed by differentiation into either chondrocytes or osteoblasts depending on the mode of ossification (endochondral or intramembranous). The individual bones of the skull then derive from one or (usually) more of these condensations. This molecular machinery is common to all condensations. However, the size of particular condensations and their subsequent rates of growth, before, during and after condensations, are important steps during which the size and shape of individual bones are determined as well as the

species-specific morphology of individual bones (Hall 2005). Genetic variation in the regulation of these processes would produce condensation specific variation that would produce covariation within regions derived from particular condensations or among regions that share regulatory influences. This is the level at which the Atchley-Hall model for developmental units or modularity in the mouse mandible was formulated (Atchley & Hall 1991), which has led to considerable work on the covariation of this complex structure (Cheverud et al 2004, Ehrich et al 2003, Klingenberg et al 2003, 2004, Mezey et al 2000).

E. *Cartilage growth*
The bones of the skull form via two distinct mechanisms of ossification. The chondrocranium, which consists of the basicranium and the capsules around the sensory capsules, forms via endochondral ossification. Here, the mesenchymal condensations differentiate into hyaline cartilage that subsequently undergoes a process of ossification. The viscerocranium also initially consists of cartilaginous elements that form within the branchial arches. The remainder of the skull consists of the cranial vault bones and bones that form around the viscerocranial elements in the face. In both of these cases, the mesenchymal condensations differentiate directly into osteoblasts and form bone directly via intramembranous ossification. The regulation of growth in bones from these two origins appears to be quite different. While the ossification and growth of membranous bone seems largely driven by the induction and mechanical influences from surrounding soft tissues (Opperman 2000, Spector et al 2002, Wilkie & Morriss-Kay 2001, Yu et al 2001), bones of endochondral origins display intrinsic growth, which is expressed initially in the interstitial growth of the hyaline cartilage and subsequently in the growth of the cranial synchondroses at specific joints between bones of endochondral origin (Cohen et al 1985, Kreiborg et al 1993). Cartilage growth thus has genetic regulatory influences that are shared among endochondral elements but not with elements which form through intramembranous ossification. Since the bones of the skull become physically adjacent and thus influence each others growth, this variation is transmitted from the chondrocranium to the rest of the skull (Hallgrimsson et al 2006). These influences, however, are secondary or epigenetic. The regulation of cartilage growth is thus likely to be a very important determinant of cranial covariation structure.

F. *Brain growth*
The brain is the largest organ in the skulls of most mammals and its growth is the primary driver for the growth of the cranial vault (Jiang et al 2002). The brain also exerts a lesser influence on the growth of the cranial base (Biegert

1963, Enlow 1990, Lieberman & McCarthy 1999). The growth of the brain thus exerts an influence on some skeletal elements and not others, and exerts its influence unequally on those that are affected. Variation in brain growth therefore translates into covariation among the elements that it influences and thus into cranial covariation structure.

G. Muscle–bone interactions

Epigenetic interactions between muscle and bone in the skull is frequently cited as a cause of covariation among skeletal elements (Cheverud 1982, Hallgrimsson et al 2004b, Willmore et al 2006a). The mechanical effects of muscle activity are known to influence bone growth (Herring 1993). Mechanical effects on bone growth are not absent during prenatal development (Delaere & Dhem 1999, Hall & Herring 1990, Herring & Lakars 1982), but are likely to increase substantially in importance once mastication begins. Therefore, bones that share muscle attachments, or are in regions in which the loads produced by particular activities such as mastication, will share common epigenetic influences that are produced by muscle activity. Such effects would produce covariation among structures that share influences (e.g. by particular muscles), or are similarly influenced by the dispersal of mechanical forces during activities like mastication (Zelditch et al 2006).

H. Somatic growth

A central thread in analyses of integration has been the relationship between the overall determinants of growth and the determinants of growth of particular structures within the organism (Castle 1929, Hallgrimsson et al 2002, Wright 1932). Variation in the overall control of growth should cause covariation among all structures. However, size-related covariation is much more complex than that for two reasons. One is that there are multiple determinants of overall growth and these determinants may affect the growth of individual structures unequally. The regulation of growth varies across ontogeny with growth hormone playing a more important role for somatic growth postnatally in mice while performing a variety of other developmental functions during fetal growth (Waters & Kaye 2002). The other is that many structures do not covary isometrically with overall size. In some cases, such as early ontogeny in many organisms, these relationships are nonlinear as well. The allometric relationship between individual structures and overall size creates an axis of covariation that is often a dominant feature of the covariation structure of a population. This is true for hominoid cranial variation (Frost et al 2003) as well as for inbred strains of mice (Hallgrimsson et al 2004b) and is probably common in any population that exhibits significant variation in body size. Allometric covariation produced by variation in size is thus a major determinant of cranial covariation structure.

The developmental palimpsest and the evolution of covariance

The list of covariance-generating processes above is not exhaustive. It is also an oversimplification in presenting an atomistic view of the development of a complex structure like the skull. It does, however, present a mechanistic account of the development of integration that leads to predictions about how mutations alter covariance structure and how natural selection acts on development to influence covariance. It also leads to a better understanding of the limitations of inference about development from patterns of phenotypic variation.

The discussion of the potential covariance-generating processes above suggests that developmental processes leave a succession of covariation imprints on the overall covariation structure. A key insight is that it is variation in these processes that produces covariance among the structures that they influence. No matter how important brain growth is as a determinant of neurocranial growth, if all individuals exhibit the same rate and amount of brain growth, then the influence of this factor on neurocranial growth will be the same in all individuals and the covariance generated by brain growth will be zero. If individuals in a population vary greatly in brain growth, then small and large brain sizes will be associated with low and high values for the measures of neurocranial growth that are influenced by brain growth. Thus, high variance in brain growth produces high integration among neurocranial structures. Importantly, the actual effect of brain growth on the neurocranium is the same in both cases. The difference is the magnitude of variance in brain growth. Thus, variance in covariance-generating development processes produces covariation at later developmental stages.

This insight leads to a prediction about the developmental basis for evolutionary changes in covariance structures. If variance in central developmental processes produces integration, then covariation structures evolve via changes to the variances of these covariation-generating processes. Any change in the variance of one of the processes relative to the variances of other relevant processes should alter covariance structure. Whether such changes increase or decrease morphological integration overall depends on which process is affected, whether its variance is increased or decreased and what its relative contribution to the covariance structure is in the ancestral or wild-type population. If the variance of a process that is the major determinant of covariance in the ancestral population is increased, then integration will also increase. The reason for this is that the proportion of the total variance explained by the axis of covariation (e.g. a principal component) that corresponds to this major developmental process will increase, leaving less variation to be explained by the less important factors. When a developmental process that contributes very little to the covariance structure in the ancestral population increases dramatically in variance, the opposite would occur and integration will decrease. This happens because an axis of covariation that explained only a minor portion of the covariance structure in the ancestral population is increased in its relative importance. The major determinants of covariance are still there, but the previously unimportant factor is now relatively more important, consequently distributing the total variance more evenly across a larger number of factors, producing a more disintegrated phenotypic covariance structure.

In the following section, we demonstrate aspects of this model through analyses of several mutants that influence the development of the mouse skull.

Mouse models for integration of the skull

The palimpsest model for the development of covariance structure makes several testable predictions. One is that a mutation that increases the variance of one of these key developmental processes should also alter the covariance structure. A second prediction is that, such a mutation should also increase or decrease overall integration. The direction of the change depends on the relative importance of the process in determining covariance structure. To test these predictions, we examined the effect of three mutations, each of which perturb one of the hypothesized processes on phenotypic variance and covariance structure.

A/WySn mice and the outgrowth of the facial prominences

A/WySn mice are homozygous for two mutations that interact epistatically to produce an elevated risk for failure of fusion of the primary palate or cleft lip (Juriloff et al 2001). Although the development of filopodial attachments at the epithelial is reduced in these mice (Forbes et al 1989), the primary morphological finding is that the outgrowth of the facial prominences is reduced or delayed during the formation of the face (Wang & Diewert 1992, Wang et al 1995). One of these factors is known to be a mutation of the *Wnt9b* gene while the other remains unknown (Juriloff et al 2006). *Wnt9b* is expressed in the facial mesenchyme and is strongly implicated in the regulation of mesenchymal proliferation (Jiang et al 2006).

In A/WySn mice, at least one of the mutations that increases cleft lip susceptibility does so by perturbing the outgrowth of the facial prominences prior and during their fusion to form the primary palate. We thus asked the question of whether variability of facial process outgrowth is increased in these mice and, if so, whether the adult covariance structure is significantly altered. As we do not yet have the two loci segregating on an A/WySn genetic background, we compared A/WySn mice to C57BL/6J mice, another inbred strain that does not develop cleft lip and has been used as the comparison in prior studies of these mice (Hallgrimsson et al 2004a, Juriloff et al 2001, Wang & Diewert 1992, Wang et al 1995).

To determine whether the variation in the outgrowth of the facial prominences is increased in A/WySn mice, we performed morphometric analyses on mouse embryos aged 9.5 to 12.0 days post-conception. We collected embryos ($n = 82$ for A/WySn and $n = 50$ for C57BL/6J), fixed them in Bouin's solution and obtained 3D computed microtomography scans of each embryo. From surface renders of the microCT data, we digitized 52 3D landmarks shown in Plate 2C and subjected these to Procrustes superimposition. All embryos were staged by counting tail somites. Since the embryos used in this study span a period during which shape changes significantly, we standardized the Procrustes data to the 16 tail somite stage. Shape variances were compared using Levene's test for the Procrustes distance as described previously (Hallgrimsson et al 2006).

As shown in Fig. 1A–B, the comparison of A/WySn and C57BL/6J embryos reveals a significant difference in shape. A/WySn embryos exhibit deficient outgrowth of the maxillary and frontonasal prominences. Our results are consistent with those reported for a 2D comparison of multiple clefting and non-clefting strains which show that A/WySn are delayed in the shape progression of the face relative to both age and tail somite stage. More importantly, the comparison of shape variance by Levene's test reveals that the outgrowth of the facial prominences in A/WySn mice is also significantly more variable (Fig. 1C).

Based on the increase in the variance of facial prominence outgrowth in A/WySn mice, the palimpsest model would predict that the covariance structure of adult A/WySn mice would be significantly altered and that the overall level of integration should be altered as well. As reported previously (Hallgrimsson et al 2004a), young adult A/WySn mice show a dramatically different covariation structure than C57Bl/6J mice (Fig. 2). The matrix correlation between the strains is only 0.24, which is lower than between related species of primates (Marroig & Cheverud 2001) and well below the resampled 95% confidence intervals of the matrices for the two strains. Morphological integration is much lower in A/WySn mice than in C57BL/6J mice as determined by resampling the scaled variance of the eigenvalues (Wagner 1989, Willmore et al 2006b) (Fig. 2B). Interestingly, the relationship between cranial shape and size is significantly reduced (Fig. 2C). We don't know how important a determinant outgrowth and fusion of the facial prominences is of the covariation structure of the mouse skull. However, this finding of a less integrated adult covariance structure would be expected if this process normally contributes very little to integration but is increased in its relative importance in A/WySn mice due to the elevated variance of facial prominence outgrowth.

The brachymorph mutation and cartilage growth

The brachymorph (bm) phenotype results from an autosomal recessive mutation in the phosphoadenosine phosphosulfate synthetase 2 gene (*Papss2*) (Kurima et al

A. Plot of PC1 against PC2 for stage-standardized Procrustes Data

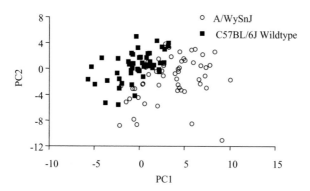

B. Wireframe deformation along PC1

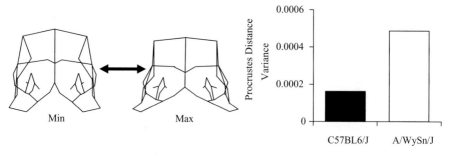

C. Variances of Craniofacial Shape

FIG. 1.   Comparison of cranial mean shape and variance for shape for A/WySn and C57BL/6J embryos aged GD 9.5–12 days. The Procrustes data were standardized to the 16 tail somite stage. (A) Plot of PC2 against PC1 for PCA analysis based on the stage-standardized data. The groups are significantly different in shape by Goodall's F-test ($P < 0.001$). (B) Wireframe deformation along the 1st PC which captures much of the difference between the two groups. (C) The shape variances for the two groups measured as the variances of the Procrustes distances from the within-group mean. The observed difference is 0.000324. The 95% upper bound on the observed difference is 0.000162. The two variances are significantly different at $P < 0.000625$ (1600 permutations).

1998). This gene codes for an enzyme involved in the sulfation pathway, catalysing the synthesis of adenosine-phosphosulfate (APS) and the subsequent phosphoryla-tion of APS to produce phosphoadenosine-phosphosulfate (PAPS). Within the Golgi apparatus, PAPS is a sulfate donor for a variety of proteins including gly-cosaminoglycans (GAGs). The mutation is a base pair substitution that causes a Gly to Arg substitution within the APS-kinase domain of the protein, and thus interferes with the second of the two functions of the enzyme (Kurima et al 1998). The result is a dramatic decrease in the availability of PAPS and thus in the sul-fation of GAGs (Orkin et al 1976). The cartilage extracellular matrix of *bm/bm* mice is therefore affected in that GAGs are undersulfated and therefore less negatively

A. Plot of PC1 against PC2 for A/WySnJ, C57BL/6J and their F1 cross.

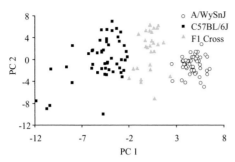

B. Integration of Craniofacial Shape at 30 days   C. Regressions of cranial shape on size in young Mice (30 days)

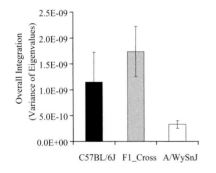

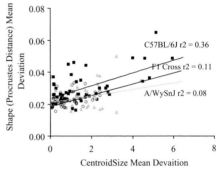

FIG. 2.   Comparisons of adult A/WySnJ mice to C57BL/6J mice at 30 days of age. (A) PCA plot comparing the two strains with their F1 cross. The groups are significantly different in shape using Goodall's F-test ($P < 0.01$). (B) Regressions of craniofacial shape (the Procrustes distance from the within group mean shape) against cranial size (centroid size). A lower proportion of the total variance in shape is explained by size in A/WySnJ mice. (C) Craniofacial integration showing that A/WySnJ mice have reduced craniofacial integration.

charged, making the proteoglycan aggregates smaller in size and less abundant (Orkin et al 1976). As a result of this altered extracellular matrix, the growth of cartilage is dramatically reduced; all skeletal elements that depend on cartilage growth are abnormally small. In the skull, the direct effects of the mutation should be confined to the chondrocranium because the growth of dermatocranial elements does not depend directly on cartilage growth (Kaufman & Bard 1999). The palimpsest model therefore predicts that if the variance of cartilage growth is increased by this mutation, the covariation structure should be altered and, depending on the importance of this process in determining covariation structure, the level of overall morphological integration should be altered as well.

We do not have a direct measure of the variance of cartilage growth for these mice. However, we have measured the overall shape variance in an adult sample

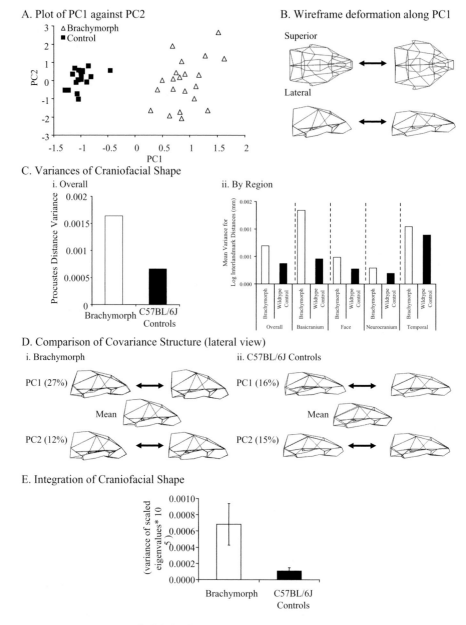

FIG. 3.   Comparisons of adult *brachymorph* homozygotes to wild-type C57BL/6J mice at >90 days of age. (A) PCA plot comparing the two groups. (B) Wireframe deformation representing the shape change along PC1. (C,i) Comparing variances for overall craniofacial shape for the two groups; (C,ii) broken down by cranial region. (D) Comparison of the significantly different covariance structures of the two groups. (E) Comparing morphological integration as measured by the variances of the eigenvalues for the two groups.

of bm/bm homozygotes on a C57BL/6J background ($n = 21$) and a sample of C57BL/6J controls ($n = 19$) (Hallgrimsson et al 2006) (Fig. 3A). All individuals were over 90 days in age. Analysis of this sample shows that the brachymorph mutation is associated with a dramatic increase in morphological variance (Fig. 3B); the Procrustes mean deviations, after adjustment for sex differences within groups, are significantly higher in the brachymorph group compared to the wild-type whether the statistical analysis is based on Levene's test (df = 39, F = 29, $P < 0.001$) or the permutation test (observed $\Delta V = 0.00097$, 1600 permutations, $P < 0.001$). Further, the largest difference in shape variances was found in cranial regions of endochondral origin (Hallgrimsson et al 2006) (Fig. 3C).

A comparison of the variance covariance matrices for the two groups reveals that the bm mutation radically alters covariance structure. The matrix correlation between the two matrices is only 0.33 (Fig. 3D). Interestingly, morphological integration is dramatically increased as determined by resampling the scaled variance of the eigenvalues (Fig. 3E). Coupled with the increased variance for shape of the endochondral portion of the skull, this result would be predicted by the model if the growth of the endochondral portion of the skull is normally a large contributor to the overall covariance structure. The increase in the variance of this process thus increases the already high proportion of the variance that is explained by this axis of covariation.

The mceph mutation and brain growth

Homozygous *mceph* (megencephaly) mutants have expanded but normally shaped brains from generalized neural cell hypertrophy generated by a single recessive autosomal mutation, an 11 bp deletion in the *Kcna1* gene (Diez et al 2003, Petersson et al 2003). Brain size is normal at birth, but increases more rapidly than in the wild-type and is increased in our sample by 30% on average by 90 days of age. Since brain growth is one of the main hypothesized integrating processes in the cranium, we predicted that if the variance of brain growth is also increased in these mice, the covariance structure will be altered. Moreover, if brain growth is a major determinant of covariation structure, overall integration will be increased.

We tested these predictions using a sample of mice carrying the *mceph* mutation on a mixed C57BL/6J*Balbc/ByJ background obtained from the Jackson Laboratory. We obtained heterozygote * homozygote mating pairs. F1 offspring were genotyped by PCR (primers: TTG TGT CGG TCA TGG TCA TC [forward], GCC CAG GGT AAT GAA ATA AGG [reverse]), and gel bands were sequenced for a subsample due to the small difference between fragment lengths. We obtained and scanned a sample of 26 *mceph/mecph* homozygotes and 25 heterozyogotes which we used as controls. As further outgroup controls, we compared these mice also to C57BL/6J wild-type mice of the same age ($n = 30$).

To determine changes in mean brain size and the variance of brain size, we obtained virtual endocasts from computed microtomography scans and calculated their volumes (Fig. 4A). Both the mean endocranial volumes and, importantly, the variance of endocranial volume are significantly higher in the *mceph* homozygotes than in either the heterozygote or C57BL/6J controls (ANOVA, $P < 0.01$, Levene's test ANOVA for variance, $P < 0.01$). Brain growth is thus increased and more variable in the *mceph* homozygotes.

As predicted, this increase in the variance of brain growth is also associated with a radically altered covariance structure (Fig. 4). The matrix correlation between the variance covariance matrices of the Procrustes coordinates for the *mceph* homozygote and heterozygote samples is 0.34, which is significantly below the resampled 95% confidence intervals of the matrices for the two strains. The overall phenotypic variance in *mceph* homozygotes is significantly greater than in the heterozygote and C57BL/6J wild-type samples (Levene's test ANOVA for Procrustes distance, $P < 0.01$). Finally, morphological integration is substantially and significantly increased in the *mceph* homozygotes as determined by resampling the scaled variances of the eigenvalues (Fig. 4).

Together, these results show that the *mceph* mutation alters the cranial covariation structure by increasing the variance of brain size which, because of the correlated responses to brain size throughout the skull, produces both an alteration in covariance structure and a significant increase in overall morphological integration.

Discussion and conclusion

The relationship between genetic and phenotypic variation is so complex that it is necessary to consider and analyse development in order to understand evolutionary changes. There are two major ways in which development complicates the genotype–phenotype map. One is through the modulation of the amount of variation. This is captured through the study of phenomena such as canalization or developmental stability. The other is through structuring or biasing the direction of variation. This latter aspect is captured through the study of phenomena like developmental constraints or morphological integration. The palimpsest model for the evolution of integration shows how intimately connected the modulation of the amount of variation is to the structuring of the direction of variance. Changes in covariance—or variation structure—are produced by altering the variance of developmental processes that affect structures differentially. The covariance structure of some complex morphological system like the skull is really the summed effect of the relative variances of all covariance-generating processes.

The latter statement is relevant to the use of patterns of covariation to dissect the developmental modularity in morphological structures (Goswami 2006a,

A. Plot of PC1 against PC2   B. Wireframe deformations along PC1 and PC2

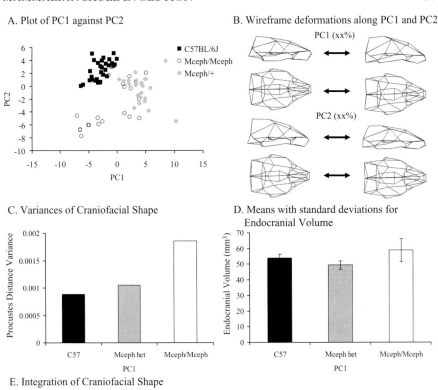

C. Variances of Craniofacial Shape   D. Means with standard deviations for
                                                                 Endocranial Volume

E. Integration of Craniofacial Shape

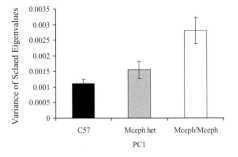

FIG. 4.   Comparisons of adult *mceph* homozygotes to heterozygotes and C57BL/6J mice at 90 days of age. (A) PCA plot comparing the three groups. (B) The wireframe deformations that correspond to the shape changes along PC1 and PC2. These components are not significantly distinct from one another and so should be interpreted as representing equivalent axes of a two dimensional plane of variation. (C) Comparing variances from craniofacial shape. The *mceph* homozygotes have significantly higher variances as determined by the delta variance permutation test in Simple 3D (Sheets 2004). (*mceph*/*mceph* vs. *mceph*/+: deltaV = 0.001, $P < 0.000625$, 1600 permutations, *mceph*/*mceph* vs C57BL/6J wild-type = deltaV = 0.00097, $P < 0.000625$, 1600 permutations). (D) Means and standard deviations for endocranial volumes for the three groups. The variances are significantly different as determined by Levene's test ANOVA ($P < 0.01$). (E) A comparison of morphological integration for the three groups. The *mceph*/*mceph* group is significantly more integrated than the other two groups as determined by resampling the variances of the scaled eigenvalues for 1000 iterations ($P < 0.001$).

Hallgrimsson et al 2002, 2004a, Magwene 2001). Wagner and colleagues have distinguished between the use of terms 'variation' and 'variability'. Accordingly, 'variation' refers to the observed sets of differences among individuals, and 'variability', to the propensity to exhibit variation (Wagner et al 1997). This is relevant here because the object of study—patterns of variability-like modularity—are properties of individual developmental systems and are products of the architecture of development. Covariation structure, however, is a property of a population that is influenced by the relative variances of covariance generating processes. As the results presented above show, the covariation structure of a sample can be radically altered by a single mutation. Yet, the architecture of individual developmental systems is the same in the sense that the potential interactions that would generate a covariance structure in a natural population are all there. The potential interaction between brain size and basicranial shape are presumably present in all the strains analysed here, but this is the dominant determinant of covariation only in the *mceph* homozygotes because of the increased variance in brain growth in this group.

This is not to say that patterns of covariation are uninformative about underlying developmental relationships. A striking result that comes out of comparative analyses of covariation structure in related species is how conservative covariation structure can be (Goswami 2006b, Marroig & Cheverud 2001, Young 2004, Young & Hallgrimsson 2005). The palimpsest model and our analysis of how specific mutations influence integration patterns, however, tells us that this conservation of integration patterns across related species is not due to the conservation of the developmental interactions that produce the covariation patterns. Covariation structure can change while those interactions remain the same. Instead, it shows that the variances of those covariance-generating processes are remarkably constant in terms of their relative magnitudes across species. So, when cranial covariation structure is similar, this means that variation in endochondral growth or brain growth are contributing equally to the covariation structures of those populations.

While our finding that single mutations can radically alter covariation structure in inbred mice is informative in terms of how covariation structure is developmentally constituted, this does not mean that single mutations are likely to produce this result in natural populations. The reasons for this are fairly obvious. The mice used here are inbred and thus have very low genetic variances. Although there is significant among-individual variation in these strains, variances for covariance-generating processes are likely to be much lower than they would be in a natural population. For this reason, relatively small changes in the variance of some covariance-generating process are needed to alter overall covariation structure in inbred mice. The changes in variance reported here are actually quite large and so produce massive changes to covariation. These mutations, with the exception of the clf1 and clf2 mutations in A/WySn mice, also produce large shifts in the phenotypic mean. Mutations of large phenotypic effect are almost always highly deleterious and thus are very unlikely to occur with sufficiently high frequency in natural populations to produce the massive shifts in covariation structure that we report here.

These results, however, are relevant for understanding the evolution of covariance in natural populations. They imply that when a change in covariance structure is seen, this means that the proportional magnitudes of the variances of the processes producing that covariance structure have been altered. So, if the covariation structure of the scapula is altered in a species with an unusually large deltoid muscle, this change probably reflects the relatively increased integrating influence of the deltoid muscle. When the covariance structure of the skull is altered in dwarfed forms, this may reflect the relatively larger influence of brain growth because of the allometric relationship between brain and body size. By arguing that the evolution of covariance thus reflects increases or decreases in the variances of integrating processes, we are making an argument that is very much in the spirit of Jacob's vision of 'evolution by tinkering' (Jacob 1977). In most cases, the evolution of covariance takes place through tinkering with the relative variances of processes that are already present.

This cannot always be the case, though. The appearance of novel structures or processes can also alter covariance. For example, a change in patterning might divide a block of tissue or some other developmental module into regions with distinct developmental-genetic influences. Once this happens, subsequent changes might build on the initial patterning change and add to the variation that is specific to each region. When existing structures such as segments or appendages are duplicated and subsequently diverge, then the covariance matrix of interest changes to include the novel structures. Within the matrix, new influences that are particular to the novel structure will alter the matrix as a whole. As in other areas, the subject of evolutionary novelty creates a set of distinct and interesting questions (Muller & Newman 2005) that are beyond the scope of this paper. The implication of the palimpsest model for how evolutionary novelty alters covariance structure is simply that such events produce new covariance-generating processes and that their relative variances will determine their influence on the covariance structure of interest.

An important implication of the palimpsest model for cranial covariation is the power of focusing on key developmental process level determinants of phenotypic variability. Although the development of the mammalian skull is very complex, we argue here and elsewhere (Hallgrimsson et al 2007) that significant axes of covariation are generated by epigenetic interactions among components of the skull. We argue that increasing or decreasing the amount of cartilage growth would act on a particular axis of morphological covariation. Similarly, increasing or decreasing the size of the brain would produce a particular axis of covariation. A full test of this proposition would compare different mutations that affect a particular process against the prediction that all would produce variation along the same axis of covariation.

If correct, our suggestion that cranial variation is driven by a finite set of key processes provides a framework with which to cut through the overwhelming complexity of head development to arrive at an understanding of how the expression of variation in the skull is structured by development. For instance, we can ask how the growth of the brain influences the shape of the skull. To understand that relationship, it may not matter how the brain increases in size. This undoubtedly matters in terms of the evolutionary explanation for increased brain size, but it probably does not matter for understanding the shape consequences of increased brain size for the skull. Similarly, the overall growth of cartilage in the skull can be regulated through many developmental pathways, each of which in turn is influenced by multiple genes. In terms of understanding how the skull changes shape, the relevant level of explanation is the interaction between chondrocranial growth and the rest of the skull. Similar arguments could be made about all of the developmental processes discussed above, as well as some others that are not discussed here. We therefore argue that a focus on key developmental processes underlying the development of a complex structure provides a framework for understanding how development structures the expression of phenotype.

Acknowledgements

This work was supported by National Science and Engineering grant 238992–02, Canadian Foundation for Innovation grant #3923, Alberta Innovation and Science grant #URSI-01-103-RI, Canadian Institutes of Health grant 131625 and Genome Canada grant (to B.H.), American School of Prehistoric Research (DEL).

References

Atchley WR, Hall BK 1991 A model for development and evolution of complex morphological structures. Biol Rev 66:101–157
Biegert J 1963 The evaluation of characteristics of the skull, hands and feet for primate taxonomy. In: SL Washburn (ed) Classification and human evolution. Chicago, Aldine, p 116–145
Brugmann SA, Kim J, Helms JA 2006 Looking different: Understanding diversity in facial form. Am J Med Genet A 140:2521–2529
Castle WE 1929 The further study of size inheritance with special reference to the existence of genes for size characters. J Exp Zool 53:421–454
Cheverud JM 1982 Phenotypic, genetic, and environmental integration in the cranium. Evolution 36:499–516
Cheverud JM 1988 A Comparison of Genetic and Phenotypic Correlations. Evolution 42:958–968

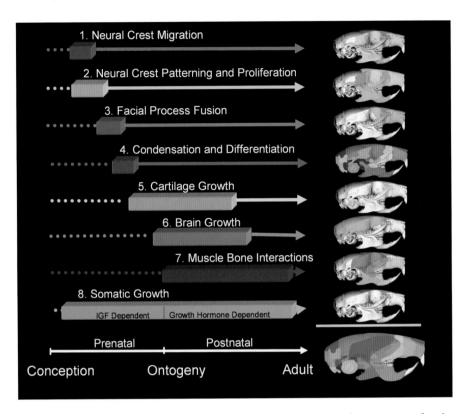

PLATE 1. Summary of the major hypothesized covariance-generating processes for the mammalian skull.

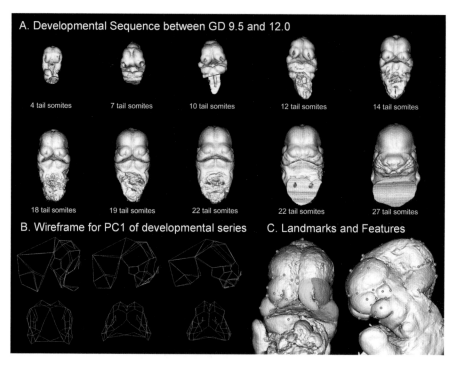

PLATE 2. The formation of the face. (A) Cross-sectional developmental sequence showing C57BL/6J embryos of different developmental stages spanning the period of the formation of the primary palate by outgrowth and fusion of the facial prominences. (B) The first principal component for a PCA analysis of Procrustes superimposed landmarks for an embryo dataset spanning this same period. (C) The landmarks used for that analysis as well as the locations of the prominences that form the primary palate. The medial nasal prominence is shaded in yellow, while the lateral nasal is red and the maxillary is in teal. All of these images and the morphometric analysis shown are based on computed microtomography scans of Bouin's fixed embryos.

Cheverud JM 1995 Morphological integration in the saddle-back tamarin (Saguinus fuscicollis) cranium. Am Nat 145:63–89

Cheverud JM 1996 Developmental integration and the evolution of pleiotropy. Am Zool 36:44–50

Cheverud JM, Ehrich TH, Vaughn TT, Koreishi SF, Linsey RB, Pletscher LS 2004 Pleiotropic effects on mandibular morphology II: differential epistasis and genetic variation in morphological integration. J Exp Zoolog B Mol Dev Evol 302:424–435

Cohen MM Jr, Walker GF, Phillips C 1985 A morphometric analysis of the craniofacial configuration in achondroplasia. J Craniofac Genet Dev Biol Suppl 1:139–165

Delaere O, Dhem A 1999 Prenatal development of the human pelvis and acetabulum. Acta Orthop Belg 65:255–260

Depew MJ, Lufkin T, Rubenstein JL 2002a Specification of jaw subdivisions by Dlx genes. Science 298:381–385

Depew M, Tucker A, Sharpe P 2002b Craniofacial Development. In: Rossant J, Tam P (eds) Mouse Development: Patterning, Morphogenesis, and Organigenesis. Academic Press, p 421–498

Diez M, Schweinhardt P, Petersson et al 2003 MRI and in situ hybridization reveal early disturbances in brain size and gene expression in the megencephalic (mceph/mceph) mouse. Eur J Neurosci 18:3218–3230

Ehrich TH, Vaughn TT, Koreishi SF, Linsey RB, Pletscher LS, Cheverud JM 2003 Pleiotropic effects on mandibular morphology I. Developmental morphological integration and differential dominance. J Exp Zoolog B Mol Dev Evol 296:58–79

Enlow DH 1990 Facial growth. Philadelphia, Saunders

Forbes DP, Steffek AJ, Klepacki M 1989 Reduced epithelial surface activity is related to a higher incidence of facial clefting in A/WySn mice. J Craniofac Genet Dev Biol 9:271–283

Frost SR, Marcus LF, Bookstein FL, Reddy DP, Delson E 2003 Cranial allometry, phylogeography, and systematics of large-bodied papionins (primates: Cercopithecinae) inferred from geometric morphometric analysis of landmark data. Anat Rec 275A:1048–1072

Gerhart J, Kirschner M 1997 Cells, embryos, and evolution: toward a cellular and developmental understanding of phenotypic variation and evolutionary adaptability. Blackwell Science, Malden, MA

Goswami A 2006a Cranial Modularity Shifts during Mammalian Evolution. Am Nat 168:270–280

Goswami A 2006b Morphological integration in the carnivoran skull. Evolution Int J Org Evolution 60:169–183

Hall BK 2005 Bones and cartilage: developmental and evolutionary skeletal biology. Elsevier Academic Press, Australia, San Diego, CA

Hall BK, Herring SW 1990 Paralysis and growth of the musculoskeletal system in the embryonic chick. J Morphol 206:45–56

Hall BK, Horstadius S 1988 The Neural Crest. Oxford University Press, London

Hallgrimsson B, Willmore K, Hall BK 2002 Canalization, developmental stability, and morphological integration in primate limbs. Yearb Phys Anthropol 45:131–158

Hallgrimsson B, Dorval CJ, Zelditch ML, German RZ 2004a Craniofacial variability and morphological integration in mice susceptible to cleft lip and palate. J Anat 205:501–517

Hallgrimsson B, Willmore K, Dorval C, Cooper DML 2004b Craniofacial variability and modularity in macaques and mice. J Exp Zool B Mol Dev Evol 302B:207–225

Hallgrimsson B, Brown JJY, Ford-Hutchinson AF, Sheets HD, Zelditch ML, Jirik FR 2006 The brachymorph mouse and the developmental-genetic basis for canalization and morphological integration. Evol Dev 8:61–73

Hallgrimsson B, Lieberman DE, Liu W, Ford-Hutchinson AF, Jirik FR 2007 Epigenetic Interactions and the Structure of Phenotypic Variation in the Cranium. Evol Dev 9:76–91

Helms JA, Cordero D, Tapadia MD 2005 New insights into craniofacial morphogenesis. Development 132:851–861

Herring S 1993 Formation of the vertebrate face: Epigenetic and functional influences. Am Zool 33:472–483

Herring SW, Lakars TC 1982 Craniofacial development in the absence of muscle contraction. J Craniofac Genet Dev Biol 1:341–357

Jacob F 1977 Evolution and tinkering. Science 196:1161–1166

Jiang R, Bush JO, Lidral AC 2006 Development of the upper lip: morphogenetic and molecular mechanisms. Dev Dyn 235:1152–1166

Jiang X, Iseki S, Maxson RE, Sucov HM, Morriss-Kay GM 2002 Tissue origins and interactions in the mammalian skull vault. Dev Biol 241:106–116

Juriloff DM, Harris MJ, Brown CJ 2001 Unravelling the complex genetics of cleft lip in the mouse model. Mamm Genome 12:426–435

Juriloff DM, Harris MJ, McMahon AP, Carroll TJ, Lidral AC 2006 Wnt9b is the mutated gene involved in multifactorial nonsyndromic cleft lip with or without cleft palate in A/WySn mice, as confirmed by a genetic complementation test. Birth Defects Res A Clin Mol Teratol 76:574–579

Kaufman MH, Bard JBL 1999 The anatomical basis of mouse development. Academic Press, San Diego

Klingenberg CP, Zaklan SD 2000 Morphological integration between developmental compartments in the Drosophila wing. Evolution 54:1273–1285

Klingenberg CP, Mebus K, Auffray JC 2003 Developmental integration in a complex morphological structure: how distinct are the modules in the mouse mandible? Evol Dev 5:522–531

Klingenberg CP, Leamy LJ, Cheverud JM 2004 Integration and modularity of quantitative trait locus effects on geometric shape in the mouse mandible. Genetics 166:1909–1921

Kraus P, Lufkin T 2006 Dlx homeobox gene control of mammalian limb and craniofacial development. Am J Med Genet A 140:1366–1374

Kreiborg S, Marsh JL, Cohen MM Jr et al 1993 Comparative three-dimensional analysis of CT-scans of the calvaria and cranial base in Apert and Crouzon syndromes. J Craniomaxillofac Surg 21:181–188

Kurima K, Warman ML, Krishnan S et al 1998 A member of a family of sulfate-activating enzymes causes murine brachymorphism. Proc Natl Acad Sci USA 95:8681–8685

Lieberman DE, McCarthy RC 1999 The ontogeny of cranial base angulation in humans and chimpanzees and its implications for reconstructing pharyngeal dimensions. J Hum Evol 36:487–517

Magwene PM 2001 New tools for studying integration and modularity. Evolution 55:1734–1745

Marroig G, Cheverud JM 2001 A comparison of phenotypic variation and covariation patterns and the role of phylogeny, ecology, and ontogeny during cranial evolution of new world monkeys. Evolution 55:2576–2600

Mezey JG, Cheverud JM, Wagner GP 2000 Is the genotype-phenotype map modular? A statistical approach using mouse quantitative trait loci data. Genetics 156:305–311

Muller GB, Newman SA 2005 The innovation triad: an EvoDevo agenda. J Exp Zoolog B Mol Dev Evol 304:487–503

Noden DM, Trainor PA 2005 Relations and interactions between cranial mesoderm and neural crest populations. J Anat 207:575–601

Olson EC, Miller RA 1958 Morphological Integration. University of Chicago Press, Chicago

Opperman LA 2000 Cranial sutures as intramembranous bone growth sites. Dev Dyn 219:472–485

Orkin RW, Pratt RM, Martin GR 1976 Undersulfated chondroitin sulfate in the cartilage matrix of brachymorphic mice. Dev Biol 50:82–94

Petersson S, Persson AS, Johansen JE et al 2003 Truncation of the Shaker-like voltage-gated potassium channel, Kv1.1, causes megencephaly. Eur J Neurosci 18:3231–3240

Raff RA 1996 The Shape of Life. University of Chicago Press, Chicago

Schlosser G, Wagner GP 2003 Introduction: The Modularity Concept in Developmental and Evolutionary Biology. In: Schlosser G, Wagner GP (eds) Modularity in Development and Evolution. University of Chicago Press, Chicago, p 1–11

Sheets HD 2004 IMP Simple3D (Software)

Spector JA, Greenwald JA, Warren SM, et al 2002 Dura mater biology: autocrine and paracrine effects of fibroblast growth factor 2. Plast Reconstr Surg 109:645–654

Wagner GP 1989 A comparative study of morphological integration in Apis mellifera (Insecta, Hymenoptera). Z zool Syst Evolut Forsch 28:48–61

Wagner GP 1995 Adaptation and the modular design of organisms. In: Morán F, Merelo JJ, Chacón P (eds) Advances in Artificial Life. Springer Verlag, Berlin, p 317–328

Wagner GP 1996 Homologues, natural kinds and the evolution of modularity. Am Zool 36:36–43

Wagner GP, Booth G, Bagheri-Chaichian H 1997 A population genetic theory of canalization. Evolution 51:329–347

Wagner GP, Mezey JG 2003 The role of genetic architecture constraints in the origin of variational modularity. In: Schlosser G, Wanger GP (eds) Modularity in Development and Evolution. University of Chicago Press, Chicago, p 338–358

Wang K-Y, Diewert VM 1992 A morphometric analysis of craniofacial growth in cleft lip and noncleft mice. J Craniofac Genet Dev Biol 12:141–154

Wang KY, Juriloff DM, Diewert VM 1995 Deficient and delayed primary palatal fusion and mesenchymal bridge formation in cleft lip-liable strains of mice. J Craniofac Genet Dev Biol 15:99–116

Waters MJ, Kaye PL 2002 The role of growth hormone in fetal development. Growth Horm IGF Res 12:137–146

Wilkie AO, Morriss-Kay GM 2001 Genetics of craniofacial development and malformation. Nat Rev Genet 2:458–468

Willmore KE, Leamy L, Hallgrimsson B 2006a Effects of developmental and functional interactions on mouse cranial variability through late ontogeny. Evol Dev 8:550–567

Willmore KE, Zelditch ML, Young N, Ah-Seng A, Lozanoff S, Hallgrimsson B 2006b Canalization and developmental stability in the brachyrrhine mouse. J Anat 208:361–372

Wright S 1932 General, group and special size factors. Genet Med 15:603–619

Young N 2004 Modularity and integration in the hominoid scapula. J Exp Zoolog B Mol Dev Evol 302:226–240

Young NM, Hallgrimsson B 2005 Serial homology and the evolution of mammalian limb covariation structure. Evolution 59:2691–2704

Yu JC, Lucas JH, Fryberg K, Borke JL 2001 Extrinsic tension results in FGF-2 release, membrane permeability change, and intracellular Ca++ increase in immature cranial sutures. J Craniofac Surg 12:391–398

Zelditch ML, Mezey J, Sheets HD, Lundrigan BL, Garland T Jr 2006 Developmental regulation of skull morphology II: ontogenetic dynamics of covariance. Evol Dev 8:46–60

DISCUSSION

Weiss: You said that if the cranium is bigger you get more flexion. This is a finding within species. Are you also saying that there is a constraint based on their evolutionary history that they have a certain basic shape? There could be multiple pathways to get the same basic shape: this must have been constrained in some way so that the alternatives for each of the components (brain and face) don't end up each with a viable structure but that doesn't 'fit' the other structure. The traits can't evolve totally independently so they are compatible with each other. Yet your analysis shows them to vary, in your experimental animals, independently. Can you clarify how you think that happens?

Hallgrimsson: The argument I am making is about the appropriate level of evolutionary explanation if we want to know the developmental determinants of craniofacial shape. If you are asking why the brain is bigger, there is a different set of answers. But if you are asking why the skull is shaped the way it is in animals that have large brains, then what matters is the developmental interaction that increased brain size produces, not a specific developmental determinant that will increase brain size. There is a level of epigenetic interaction there which we understand: the relationship between brain size and the shape of the skull. The lower levels are interesting but we can forget about them for the purposes of that particular explanation.

Lieberman: These components are already deeply constrained by the general structure of skulls. Brains have to sit on cranial bases and inside the neurocranium, and faces have to grow forwards. These architectural spatial relationships constrain the way in which bones grow, limiting the number of ways these units interact in the first place. The system constrains how variation can interact.

Cheverud: The constraint is even directional: the cranial base will be formed before the face. It is not a two-way thing. The face has to adjust.

Budd: Why did you choose to call these three components of the skull 'modules'?

Hallgrimsson: They are modules by virtue of the dominant influences of particular developmental processes. For instance, the basicranium is primarily endochondral in origin.

Budd: So it is a developmental definition.

Hallgrimsson: We are making assumptions about the major developmental influences acting on particular regions of the skull.

Lieberman: There is also a classic paper showing that in terms of the various structures of the skull, these units do actually behave as units (Chevrud 1982).

Cheverud: This was drawn from the work of Melvyn Moss. If you modify the growth of the brain it has a massive effect on the brain case and very little effect on the face. Various other ablation experiments and studies of human disease showed that there was a modular reaction to the change in different cranial organs.

Coates: You appear to have the data to look at other possible modules where you might expect to find separation of variation, such as within the mandibular arch, or derivatives of other visceral arches.

Hallgrimsson: This is a difficult thing to do. We have done that in outbred mice and have tried to replicate Jim Chevrud's studies on primates. The covariation structure of the mouse skull seems to be quite different from primates, however. We detect the chondrocranial module, but the base and neurocranium don't separate out so much. The problem is the number of potential modular developmental processes. The importance of each of these processes in different species can vary, which would produce a different covariance. If we see similarity in covariance structure this is not because these developmental systems are not capable of producing different covariance structures, it will probably because the variance associated with the difference tends to be similar across species.

Hanken: Even the most ardent selectionist wouldn't deny that in any generation variation is limited. This in itself doesn't have any significance for evolution. It is whether over evolutionary time the developmental system will still limit variation in a predictable way despite strong selection to do otherwise. Yesterday we learned that the covariance matrix itself is subject to selection. Where is there great significance here for microevolution or macroevolution, given that it is artificial strains looking over short numbers of generations, and the features themselves are subject to selection?

Cheverud: The alleles that we are sampling the effects of are drawn from the base population of all mice, which was a mixed population of different subspecies. They are natural alleles that are drawn from populations. The effects of the alleles aren't so unnatural; the unnaturalness of the population is in its allele frequencies, designed to be powerful for gene mapping. The gene frequencies are ideal for mapping.

Stern: I would take issue with the statement that the alleles are 'natural'. They are alleles that survive an inbreeding process in a laboratory. We have no idea whether that represents a biased subset of alleles.

Lieberman: True enough. But one idea behind it was not to look at particular alleles, but rather to try to do an experiment in which we modified the relative size and shape of a particular skull. The only other way people have tended to do this is to do general comparative analyses of different organisms. The problem with that type of analysis is you don't know what all the other effects are that are taking place.

Hallgrimsson: The mouse models we are using don't give the complete answers. It is one way of approaching the question. The advantage we have is that we know what the developmental process is. If you want to do this in an outbred mouse you would come up with some interesting morphometric results, but they would be hard to interpret. The relationship between the growth of the brain and the length of face and basicranium is an interesting one, because this can be tested.

Wilkins: I was intrigued by your statement that the developmental architecture remained the same. I would say that developmental processes remain the same on the whole, but the genetic architecture is presumably subtly different in these mutants. Would you agree with this distinction?

Hallgrimsson: Yes.

Brakefield: Do you think you would see a lot of mutant–environment interactions if you did the whole study in a different setting? Presumably these animals are reared in a favourable environment. Would you see different patterns if you were to rear the same series of mutants in different temperatures or a more natural diet?

Cheverud: In the US that would probably be illegal.

Lieberman: Someone needs to release some C57BL onto an island somewhere!

Brakefield: How robust is your interpretation of phenomena such as integration if you also take an environmental axis into account?

Hallgrimsson: If you introduce environmental variance, you are introducing covariation.

Jernvall: This is something that we have seen with teeth. Every time there is a knockout or a mouse with experimentally altered level of gene expression, variance goes through the roof. This raises the question if you keep breeding these mutant strains does the variance go down?

Hallgrimsson: No one has done this.

Hall: With the *brachymorph* mutant you said the integration was much higher. I assumed this was because *brachymorph* had been inbred for some time.

Hallgrimsson: The brachymorphs and the C57 controls should differ only in the presence of the *brachymorph* mutation. The same is true for the other models in which the comparison is to littermate siblings that are not carrying the mutation. So there should be no difference in inbreeding between the mutants and controls in any of the studies that I talked about.

Oxnard: Our morphometric studies were with primates. We were looking at particular areas of the skull using large numbers of morphometric points. The first time that we did this we identified exactly the units that you mentioned and we started to set up hypotheses about subunits of those areas. Then we found that if we took the mandible and separated the tooth bearing and the non-tooth bearing parts, and then did the same with the maxilla, we got to a point where we had to include part of the cranial base. This comprises the basi-sphenoid and the basi-occiput. Though the initial studies recognized this all as one unit, the cranial base, the later studies implied something different. They implied that changes were occurring not in the two bony parts alone, but in the conglomeration comprising all the centres of ossification and the cartilaginous partitions between them. There are as many as eight centres of ossification involved here. As we go from front to back we can see changes in these units going from associations with the face to associations with the cranium. This again makes me wonder about the reality of units, and whether or not there is a different kind of integration going on here.

Hallgrimsson: That was the point of my preface in which I was superimposing these different developmental processes. There is a large number of developmental processes that leave covariation signatures. All you need is for a process to affect some structures and not others for this to influence the covariation structure. Since you have many of these in the skull, there is a complex series of overlaying covariation structures. When you are looking at the adult covariation structure, whether or not you can pick up on these depends on how strong they are.

Oxnard: I am talking about primates, but it seems they can be picked up strongly once you look for them. If you assume you have a module you won't find continuity.

Hallgrimsson: The corollary to that is when you do this sort of analysis you get an uninterpretable matrix in which there doesn't appear to be a pattern. It doesn't mean development isn't modular. It means not enough of the modules are important enough to leave a strong enough imprint.

Oxnard: This leads to something else. One has to use anatomy to define points to do morphometrics. People argue about these points. For example, the base and apex of the styloid process are commonly used as such points. The styloid process, however, comprises as many as four centres of ossification. Which centres are actually fused into the final adult styloid process is quite variable. Therefore the points based upon it may well not be developmentally homologous. Such points may be different things in different animals.

Lieberman: Earlier you described the general structure and development of the mammalian skull. One interesting point is that the kinds of covariance interactions seen in the mice are surprisingly similar to the ones seen across humans. This tells us that the basic construction of the skull is more-or-less the same in different creatures. They are altered in similar ways leading to only so many kinds of variation.

R Raff: I speak from total ignorance on this topic. Supposing you select for horns, and get horned mice. What is likely to happen? Will you be breaking some constraints?

Hallgrimsson: Then you have introduced a new covariance determining process. Chris Klingenberg and one of his students have done a morphometric analysis of sheep with and without horns. The covariance structure is dominated by the horns if they are there.

Ackermann: Part of the point in getting an alteration in the covariance structure is that this may itself be an indicator of selection. Having a relatively stable covariance structure across early development, or changes in covariation early in development, might be indicative of the action of non-random process. You should be able to test for this.

Lieberman: It would be interesting to look at these covariation structures over ontogeny.

Ackermann: You would have to do it on a population level.

Cheverud: There are very few such populations around, but there are some.

Reference

Cheverud J 1982 Phenotypic, genetic, and environmental morphological integration in the cranium. Evolution 36:499–516

The developmental genetics
of microevolution

David L. Stern

Department of Ecology & Evolutionary Biology, Princeton University, Princeton, NJ 08544, USA

Abstract. What is the relationship between variation that segregates within natural populations and the differences that distinguish species? Many studies over the past century have demonstrated that most of the genetic variation within natural populations that contributes to quantitative traits causes relatively small phenotypic effects. In contrast, the genetic causes of quantitative differences between species are at least sometimes caused by few loci of relatively large effect. In addition, most of the results from evolutionary developmental biology are often discussed as though changes at just a few important 'molecular toolbox' genes provide the key clues to morphological evolution. On the face of it, these divergent results seem incompatible and call into question the neo-Darwinian view that differences between species emerge from precisely the same kinds of variants that segregate much of the time in natural populations. One prediction from the classical model is that many different genes can evolve to generate similar phenotypes. I discuss our studies that demonstrate that similar phenotypes have evolved in multiple lineages of *Drosophila* by evolution of the same gene, *shavenbaby/ovo*. This evidence for parallel evolution suggests that *svb* occupies a privileged position in the developmental network patterning larval trichomes that makes it a favourable target of evolutionary change.

2007 Tinkering: the microevolution of development. Wiley, Chichester (Novartis Foundation Symposium 284) p 191–206

One major goal of evolutionary genetics is to determine the characteristics of the genetic changes that give rise to phenotypic differences within and between species. Does evolution occur by the fixation of mutations that cause relatively discontinuous jumps in the phenotype or by the fixation of mutations that cause more subtle changes? Both views of the mutational basis for evolution have been supported by theoretical arguments and empirical observations ever since Darwin proposed that changes of small phenotypic effect seemed more plausible. Even in Darwin's time, some of his colleagues suggested that more discontinuous events probably contributed to evolutionary change. The entire 19th century discussion was hampered, however, by incorrect notions of the basis for heredity.

The 're-discovery' of Mendelian genetics in 1900, while providing clues to the material basis of heredity, did not by itself resolve the question. The neo-Mendelians, such as William Bateson and Thomas Morgan, felt that the relatively dramatic mutations discovered in the laboratory provided evidence that mutation was the driving force of evolutionary change. At about the same time R. A. Fisher illustrated that particulate inheritance through Mendelian laws was not only compatible with Darwinian natural selection, but necessary for its proper function. He went on to claim, based on an attractively simple mathematical model, that mutations of very small fitness effect were most likely to be fixed during evolution by natural selection.

Despite the explosion of genetic and genomic approaches in the past few decades, in only a few cases have the individual genes responsible for phenotypic variants been identified. In even fewer cases have the individual nucleotides been identified. Most of the data that might be thought to be relevant comes from studies of what is called the 'genetic architecture' of variation. This usually means studies of the magnitude of effects of quantitative trait loci (QTLs) and their pleiotropic and epistatic effects (Mackay 2001). Most studies include insufficient sample sizes to provide reliable estimates of epistasis nor are they capable of detecting variants of small effect, so studies of genetic architecture usually amount to estimates of the minimum number of genomic regions contributing to variation. In addition, single QTLs of large effect can contain multiple genetic variants (Orgogozo et al 2006), so it is not yet clear how many QTLs of large effect reflect single variants of large effect.

Despite these methodological limitations, there is a general consensus in the literature that QTL studies of intraspecific variation provide information relevant to the question 'which variants lead to evolutionary change between species?' I believe this is incorrect. This idea is essentially an extrapolation of Darwin's hypothesis that the variants that can be observed in natural populations are precisely the variants that will lead to differentiation between species. There is an alternative hypothesis that can be stated most clearly in its extreme form. This alternative is that most variants segregating in natural populations are irrelevant to long-term evolution. For example, they may represent variants that are maintained by balancing selection or by mutation-selection balance, but that are never fixed. The differences between species may reflect rare adaptive mutations that are rapidly swept to fixation and are thus difficult to observe in contemporary populations. Of course, reality may sit somewhere between these extreme models and neutral variants may also contribute to phenotypic differences between species. The point is that studying only variation within species is insufficient to test between these competing models.

A second approach, therefore, is to identify the variants that distinguish species and determine whether they represent a non-random sample of potential variants. Determining the null model in this case is non-trivial. One might identify all of the genes that have the potential to generate variation in a particular phenotype, for example by a mutagenesis experiment, and then determine whether the observed natural variants represent a special subset of all possible variants. However, it is not clear that any single experiment has the ability to identify all possible molecular variants, or even an unbiased sample, that might influence a trait, particularly because some traits are controlled by genes with redundant functions.

A third approach is to identify variants differentiating species and compare them with variants segregating within species, but this approach also has shortcomings. It is not clear if the goal should be to determine whether the variants within and between species are the same or whether they are different. These alternatives imply different evolutionary models. For example, if the variants are different, then it might suggest that the variants distinguishing species are in some way special and that variants segregating within species have little long term evolutionary potential. Alternatively, this result might suggest that differences between species arise from a large collection of possible variants within species and only a small proportion have so far been fixed between species. The contrasting result, that the variants within species occur at the same loci that differentiate species, implies that only a subset of genes can *vary* to generate variation for particular phenotypes. Alternatively, it might imply that variants that are normally maintained within species, for example, by balancing selection, were fortuitously fixed.

It is clear that it is not possible to distinguish between alternative evolutionary models simply by examining the distribution of genes that vary within a species, between a species or by a comparison of within and between species variation. Possibly the only way around this problem is to study convergence, where similar phenotypes have evolved repeatedly in different lineages. One approach would be to determine whether a particular subset of genes evolves repeatedly for similar functions in different lineages. This phenomenon is called parallelism and there are now several examples of parallelism in which the individual genes have been identified (Colosimo et al 2005, Price & Bontrager 2001, Sucena et al 2003). A conceptually related approach is to determine whether multiple substitutions have occurred in a single gene to cause a particular phenotypic variant (Molla et al 1996).

The evolution of *cis*-regulatory regions

Recently, the application of molecular biology, genetics and developmental biology to evolutionary questions has led to the identification of several genes responsible for natural phenotypic variation (Colosimo et al 2005, Johanson et al 2000, Shapiro et al 2004, Sucena & Stern 2000, Werner et al 2005). In many cases, morphological changes appear to have resulted from changes in *cis*-regulatory regions of conserved genes, rather than in changes to the protein coding regions (Colosimo et al 2005, Shapiro et al 2004, Stern 1998, Sucena et al 2003, Sucena & Stern 2000). However, in only a few cases have the individual causal nucleotide changes been identified (Gompel et al 2005, Prud'homme et al 2006, Wang & Chamberlin 2002).

There are theoretical reasons for expecting *cis*-regulatory regions to be common sites of evolutionary change. For genes involved in development, mutations in *cis*-regulatory regions are expected to have fewer pleiotropic consequences than mutations in protein coding regions (Carroll et al 2001, Davidson 2001, Stern 2000, Wilkins 2002). This is because changes in *cis*-regulatory regions are thought to change only one or a few aspects of the regulation of a gene, whereas changes in a protein-coding region may alter the gene's function whenever it is expressed. *Cis*-regulatory regions therefore generate modularity within a single gene and the individual modules may be more susceptible to evolutionary change.

Regulatory modules of enhancer activity have been found for a few well-studied genes (Davidson 2001, DiLeone et al 1998, Small et al 1992, Yuh et al 1998). These modules consist of clusters of transcription factor binding sites that act together to encode a transcriptional output (Berman et al 2002, Davidson 2001, Gompel et al 2005, Markstein et al 2002, Yuh et al 2001). A single 'gene' can possess multiple enhancer modules, each coding for expression in a different cellular context (Davidson 2001).

Evolution of the *shavenbaby-ovo* gene

We have been studying the genetic basis for a superficially simple phenotypic difference between *Drosophila* species in an attempt to generate the type of data that might be useful for answering this classic problem in evolutionary biology. The island endemic species *Drosophila sechellia* has evolved the absence of microtrichiae (hereafter called trichomes) on the anterior compartment of each segment along the dorsal and lateral surface of the first-instar larvae (Fig. 1).

The function of larval dorsal trichomes is not known. Only first-instar larvae produce these dorsal and lateral trichomes; second- and third-instar larvae lose precisely the same trichomes that are absent from the *D. sechellia* first-instar larva (Kuhn et al 1992). The evolutionary loss of trichomes in first-instar larvae of *D. sechellia* may therefore represent a heterochronic shift of the pattern normally found in second-instar larvae to first-instar larvae. More than 50 *Drosophila* species that we have examined show a pattern of trichomes in dorsal and lateral regions similar to *D. melanogaster* (unpublished data), suggesting that the presence of these dorsal and lateral trichomes serves a function specifically in first-instar *Drosophila* larvae.

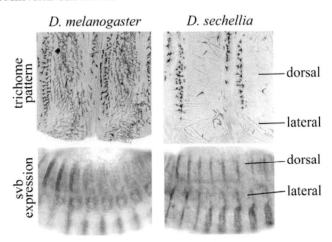

FIG. 1.   Pattern of dorsal and lateral trichomes in *D. melanogaster* and *D. sechellia* are shown in the top row. *In situ* hybridization of *svb* antisense mRNA against embryos of *D. melanogaster* and *D. sechellia* are shown in the bottom row. The dorsal and lateral domains in which trichomes and *svb* expression are lost in *D. sechellia* are indicated. Pictures of embryos showing the *svb* expression pattern were kindly provided by Isabelle Delon.

The loss of trichomes in *D. sechellia* may represent either a relaxation of this selection pressure or selection for a new function. For example, it is likely that ventral trichomes are important in locomotion (Inestrosa et al 1996) and it is possible that dorsal and lateral trichomes also aid in locomotion through some substrates. The loss of dorsal and lateral trichomes in *D. sechellia* may, therefore, be an adaptation to a specific substrate. For example, *D. sechellia* feeds on recently fallen fruits of the *Morinda citrifolia* tree (R'kha et al 1991). These recently fallen fruits produce compounds that are toxic to most *Drosophila* species, although other *Drosophila* species can utilize *Morinda* fruits after the fruits have rotted for several days and most of the toxic compounds have degraded. *D. sechellia* is resistant to the toxins produced by *Morinda* and this enables it to colonize recently fallen fruits. It is possible that recently fallen fruits provide a different challenge for *Drosophila* larvae burrowing into the flesh and that loss of trichomes aids this behaviour, but this hypothesis has not yet been tested. Despite the current uncertainty about the adaptive significance of the loss of trichomes, the patterning of trichomes has served as an important model for studying development (Lewis 1978, Nusslein-Volhard & Wieschaus 1980) and the evolution of these trichome patterns has provided a powerful model for studying the genetic basis for morphological evolution (Sucena et al 2003, Sucena & Stern 2000).

In genetic crosses, the *D. sechellia* phenotype is completely recessive to the hairy phenotype of *D. melanogaster, D. simulans*, and *D. mauritiana* (Sucena & Stern 2000).

In an initial study, we showed that the evolutionary difference in trichome pattern mapped to the *shavenbaby-ovo* (*svb*) gene and that altered expression of *svb* corresponded precisely with the change in trichome pattern (Fig. 1). The *svb* gene is a transcription factor that is both necessary and sufficient to direct trichome differentiation (Payre et al 1999). That is, if *svb* is artificially expressed in cells in which it is not normally expressed, it leads to the autonomous differentiation of trichomes. In contrast, if a dominant negative form of *svb* is expressed in cells that normally differentiate trichomes, then these cells differentiate naked cuticle.

Evidence for parallel evolution of *svb*

Examples of parallelism can provide important insights into how developmental pathways can evolve to generate novelty. In particular, these studies suggest that genes in some positions in developmental pathways are more likely to accumulate evolutionarily relevant mutations, perhaps because these mutations cause fewer deleterious pleiotropic consequences than mutations at other points in developmental pathways (Carroll et al 2001, Davidson 2001, Stern 2000, Wilkins 2002).

In addition to our study of *D. sechellia*, several studies suggest that evolution of *svb* regulation in other fly lineages has caused convergent evolution of trichome patterns. First, changes in trichome pattern similar to those observed in *D. sechellia* can also be found in the *D. virilis* species group (Dickinson et al 1993, Sucena et al 2003). The most current phylogeny of this group (Spicer & Bell 2002) suggests that the trichome pattern has evolved at least twice, and perhaps three times, independently within the *D. virilis* group (Fig. 2). We demonstrated (Sucena et al 2003) that expression of *svb* is precisely correlated with trichome patterns in these species (Fig. 2) and genetic crosses indicate that the trichome pattern segregates with the X chromosome, the location of the *svb* gene. In addition, the novel trichome pattern found in the olive fruit fly, *Bactrocera oleae*, is precisely correlated with expression of *svb* (Khila et al 2003), again suggesting that regulation of *svb* expression is related to trichome patterning. In neither of these two examples, however, has it been proven that the evolution of the trichome pattern is caused in part or in whole by evolution of *svb*.

While these studies provide some evidence that *svb* is a favoured target for evolution of larval trichome patterns, several of the caveats discussed earlier are relevant. First, it is not entirely clear which genes should serve as a null set for comparison. One possibility would be to include all genes that are known to alter trichome development and patterning. But it is not clear if this is a relevant comparison since not all of these genes may be active (or may have the potential to be active) in the first-instar larva. An alternative set would be all genes that are involved in trichome patterning and development in the particular region of the larvae where the trichomes are developing. We currently know some of the genes that are involved in

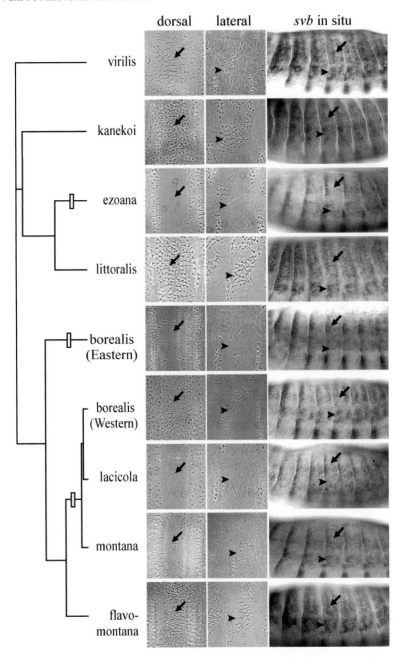

FIG. 2. Nine species of the *D. virilis* group showing a precise correlation between trichome patterns of the dorsal and lateral third abdominal segments and *svb* expression in embryonic epidermis. A molecular phylogeny is shown to the left and the dorsal (arrows) and lateral (arrowhead) trichome and *svb* expression patterns are shown in columns on the right. Figure from Sucena et al (2003) Nature 424:935–938.

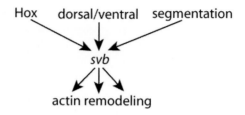

FIG. 3.   The *svb* gene sits at a regulatory nexus between 'upstream' patterning genes, such as Hox genes and segmentation genes, and 'downstream' genes involved in remodelling actin to form cuticular trichomes.

patterning trichomes on larvae, but it is likely that not all have been identified. Nonetheless, it may be worthwhile to build a simple model based on what is currently known about the regulatory cascade patterning trichomes.

The particular pattern of trichomes on first-instar larvae reflects information that derives from a variety of patterning genes, including segmentation genes and Hox genes that provide positional information along the anterior-posterior axis. It is likely that many of these genes regulate the *svb* gene, but there is currently only published evidence for the *wingless* pathway (Payre et al 1999). In addition, we do not yet know whether any of these pathways regulate *svb* directly. Nonetheless, it is clear that considerable patterning information must somehow regulate the complex expression pattern of the *svb* gene (see Figs 1,2). The *svb* gene, in turn, regulates multiple genes that reorganize the actin cytoskeleton prior to cuticle deposition (Delon et al 2003, Chanut-Delalande et al 2006).

The *svb* gene is unique in this network of genetic interactions because it sits at the nexus of the positional information from segmentation and Hox genes and the downstream genes that control actin distribution in the cell (Fig. 3). While trichome distribution *could* be changed by altering any of at least several dozen genes, *svb* can be changed so that it alters only trichome patterns without affecting segmentation or other cell biological processes (Delon & Payre 2004). It is possible that the limited pleiotropic consequences of altering *svb* expression make it a favoured target for evolving new trichome patterns.

Conclusions

Detailed study of the genetic changes that cause differences between closely related species can provide insight into fundamental questions in evolutionary biology. In particular, comparisons of closely related species are more likely to reveal the actual genetic changes that cause phenotypic divergence than comparisons between distantly related species. This is both because of methodological limitations on studying distantly related species but also because multiple substitutions will accumulate over time that may disguise the original changes. In addition, studies of variation within species suffer from the problem that we do not yet know whether all of this variation is even potentially relevant to long-term evolution or whether there are biases in which kinds of variants are favoured by selection. The existence of multiple cases of parallelism, evolution of similar phenotypes in different lineages by the same genes, suggests that some genes are favoured targets of evolutionary change and may be alerting us to the possibility that not all variants found in natural populations have an equal chance of contributing to long-term evolutionary changes.

References

Berman BP, Nibu Y, Pfeiffer BD et al 2002 Exploiting transcription factor binding site clustering to identify cis-regulatory modules involved in pattern formation in the *Drosophila* genome. Proc Natl Acad Sci USA 99:757–762

Carroll SB, Grenier JK, Weatherbee SD 2001 From DNA to diversity: Molecular genetics and the evolution of animal design. Blackwell Science, Malden

Chanut-Delalande H, Fernandes I, Roch F, Payre F, Plaza S 2006 Shavenbaby couples patterning to epidermal cell shape control. PLoS Biol 4:e290

Colosimo PF, Hosemann KE, Balabhadra S et al 2005 Widespread parallel evolution in sticklebacks by repeated fixation of Ectodysplasin alleles. Science 307:1928–1933

Davidson EH 2001 Genomic regulatory systems. Academic Press, San Diego

Delon I, Payre F 2004 Evolution of larval morphology in flies: get in shape with shavenbaby. Trends Genet 20:305–313

Delon I, Chanut-Delalande H, Payre F 2003 The Ovo/Shavenbaby transcription factor specifies actin remodelling during epidermal differentiation in Drosophila. Mech Dev 120:747–758

Dickinson WJ, Tang Y, Schuske K, Akam M 1993 Conservation of molecular prepatterns during the evolution of cuticle morphology in Drosophila larvae. Evolution 47:1396–1406

DiLeone RJ, Russell LB, Kingsley DM 1998 An extensive 3′ regulatory region controls expression of *Bmp5* in specific anatomical structures of the mouse embryo. Genetics 148:401–408

Gompel N, Prud'Homme B, Wittkopp PJ, Kassner VA, Carroll SB 2005 Chance caught on the wing: *cis*-regulatory evolution and the origin of piment patterns in *Drosophila*. Nature 433:481–487

Inestrosa NC, Sunkel CE, Arriagada J, Garrido J, Godoy-Herrera R 1996 Abnormal development of the locomotor activity in yellow larvae of Drosophila: a cuticular defect? Genetica 97:205–210

Johanson U, West J, Lister C, Michaels S, Amasino R, Dean C 2000 Molecular analysis of FRIGIDA, a major determinant of natural variation in *Arabidopsis* flowering time. Science 290:344–347

Khila A, El Haidani A, Vincent A, Payre F, Souda SI 2003 The dual function of ovo/shavenbaby in germline and epidermis differentiation is conserved between Drosophila melanogaster and the olive fruit fly Bactrocera oleae. Insect Biochem Mol Biol 33:691–699

Kuhn DT, Sawyer M, Ventimiglia J, Sprey TE 1992 Cuticle morphology changes with each larval molt in D. melanogaster. Drosoph Inf Serv 71:218–222

Lewis EB 1978 A gene complex controlling segmentation in Drosophila. Nature 276:565–570

Mackay TFC 2001 Quantitative trait loci in *Drosophila*. Nat Rev Genet 2:11–20

Markstein M, Markstein P, Markstein V, Levine MS 2002 Genome-wide analysis of clustered Dorsal binding sites identifies putative target genes in the *Drosophila* embryo. Proc Natl Acad Sci USA 99:763–768

Molla A, Korneyeva M, Gao Q et al 1996 Ordered accumulation of mutations in HIV protease confers resistance to ritonavir. Nat Med 2:760–766

Nusslein-Volhard C, Wieschaus E 1980 Mutations affecting segment number and polarity in Drosophila. Nature 287:795–801

Orgogozo V, Broman KW, Stern DL 2006 High-resolution quantitative trait locus mapping reveals sign epistasis controlling ovariole number between two Drosophila species. Genetics 173:197–205

Payre F, Vincent A, Carreno S 1999 *ovo/svb* integrates Wingless and DER pathways to control epidermis differentiation. Nature 400:271–275

Price T, Bontrager A 2001 Evolutionary genetics: the evolution of plumage patterns. Curr Biol 11:R405–408

Prud'homme B, Gompel N, Rokas A et al 2006 Repeated morphological evolution through cis-regulatory changes in a pleiotropic gene. Nature 440:1050–1053

R'kha S, Capy P, David JR 1991 Host-plant specialization in the *Drosphila melanogaster* species complex: a physiological, behavioral and genetical analysis. Proc Natl Acad Sci USA 88:1835–1839

Shapiro MD, Marks ME, Peichel CL et al 2004 Genetic and developmental basis of evolutionary pelvic reduction in threespine sticklebacks. Nature 428:717–723

Small S, Blair A, Levine M 1992 Regulation of *even-skipped* stripe 2 in the *Drosophila* embryo. EMBO J 11:4047–4057

Spicer GS, Bell CD 2002 Molecular phylogeny of the Drosophila virilis species group (Diptera: Drosophilidae) inferred from mitochondrial 12S and 16S ribosomal RNA genes. Ann Entomol Soc Am 95:165–161

Stern DL 1998 A role of *Ultrabithorax* in morphological differences between *Drosophila* species. Nature 396:463–466

Stern DL 2000 Perspective: Evolutionary developmental biology and the problem of variation. Evolution 54:1079–1091

Sucena E, Stern DL 2000 Divergence of larval morphology between *Drosophila sechellia* and its sibling species caused by cis-regulatory evolution of ovo/shaven-baby. Proc Natl Acad Sci USA 97:4530–4534

Sucena E, Delon I, Jones I, Payre F, Stern DL 2003 Regulatory evolution of shavenbaby/ovo underlies multiple cases of morphological parallelism. Nature 424:935–938

Wang X, Chamberlin HM 2002 Multiple regulatory changes contribute to the evolution of the *Caenorhabditis lin-48 ovo* gene. Genes Dev 16:2345–2349

Werner JD, Borevitz JO, Warthmann N et al 2005 Quantitative trait locus mapping and DNA array hybridization identify an FLM deletion as a cause for natural flowering-time variation. Proc Natl Acad Sci USA 102:2460–2465

Wilkins AS 2002 The evolution of developmental pathways. Sinauer Associated, Sunderland

Yuh C-H, Bolouri H, Davidson EH 1998 Genomic cis-regulatory logic: experimental and computational analysis of a sea urchin gene. Science 279:1896–1902

Yuh C-H, Bolouri H, Davidson EH 2001 Cis-regulatory logic in the endo16 gene: switching from a specification to a differentiation mode of control. Development 128:617–629

DISCUSSION

R Raff: How fluid are these enhancers? You focused on the particular phenomenon you are looking at with that hair. What if you took similar enhancers where you are not seeing a difference in phenotype? Is it possible that enhancer elements are coming in and going out, being compensated for, and so forth? In other words, if you had the microscopic structure you might find there are a lot of changes going on, even though the output at the end is trichomes, or not trichomes.

Stern: Every single one of those evolving enhancers has expression patterns that are evolutionarily conserved. Parts of the pattern are evolving. The same enhancers are coding for conserved as well as evolving bits. Different elements are clearly regulated in different ways by the Hox genes. Our goal is to identify all the binding sites for all of these expression patterns, so we can say which ones are evolving different functions, and which ones are evolving new binding sites but retaining a conserved patterning function.

Duboule: I think you could explain your point in another way. The reason you don't have the trichomes in *D. sechellia* is because there is a *cis*-located repressive activity, which is the target of your evolution. This could explain why, in your enhancer bashing experiment, you get de-repression at the end, which is hard to explain if you talk about just enhancer evolution. If you assume that in the wild-type *D. sechellia* there is the potential to express in trichomes, but because of the overall architecture of the locus you repress these enhancers. Then if you take it out and put it as a positional effect configuration, you can see this de-repression that would not normally occur in *D. melanogaster*, and then see these trichomes coming. So the question is, what about sequences? If you align all these enhancer sequences, what does it tell you?

Stern: The species are diverged by about 500 000 years. They show about 1% sequence divergence. We have these enhancers down to about 1 kb, with about 10 base pair changes. There are larger sequence differences such as big insertions, but we can exclude all of them by recombination analysis. We are therefore dealing with reasonably small regions with relatively few sequence changes. The other question of yours concerned repression. You posited a specific model of repression and de-repression. I'll propose a more general one: you can get exactly the same pattern through either a loss of activating sites or a gain of repressing sites. If it is the latter, this would explain our strange expression patterns a little better. It is hard to explain the whole thing on the basis of a single repression region because of our recombinants going in both directions.

Duboule: That should be a global chromosomal architecture, which involves both the 3′ and 5′.

Stern: It is not possible to explain our current results with less than three changes.

Duboule: We heard earlier in the meeting that what counts is not only the cell, but also the environment of the cell. This cell doesn't care about what runt is doing five cell rows ahead unless the runt gene in itself is producing something that affects the cell. Therefore I am not convinced by your 'pathwork': I think the network should be extending not only to what is happening within the cell, but also what the cell is receiving in terms of signals.

Stern: That is a pathwork of transcription factors. There are signalling molecules that will input, but the only thing the enhancer cares about is the transcription factors that are expressed. Some of those will be expressed or repressed on the basis of signals received. Of course, the external environment matters, but all the *cis*-regulatory region cares about is the transcription factors. This is what generates the modularity within enhancers, and thus their evolvability.

Lieberman: Irma Thesleff, you talked about a similar issue. Would you say the same principles would apply?

Thesleff: During development the dynamic signal networks are key regulators of the advancing differentiation of all cells, which is seen as changes in the array of transcription factors expressed in the cells. Of course, you can leave out the signals from outside, but since these signals will induce the expression of the transcription factors and change the composition of the effectors inside the cell, it is difficult for me to consider the changed composition of transcription factors separately from the signals networks.

Stern: The main point is that you can ignore most of these networks to get at what is evolutionarily relevant.

Laubichler: The pathway idea in relation to a network amounts to a decomposition problem. You have to identify invariant properties of the network in the context of a specific process that you are interested in. Then you can justify dropping off certain areas of the overall network and focusing just on one, because for this particular evolutionary scenario this is all that is relevant.

Stern: The reason it is easy to see with the *svb* network is because the output is straightforward: it is production of a trichome or not. This model clarifies the issue; the structure of that network is very important. Most developmental processes will be more complex. There are cell movements and there is complex differentiation. However, it is probably possible to decompose most or all developmental processes into simpler steps. Cells make decisions at particular times in development, and each of these decisions can be relatively independent of others. They are dependent on signals the cell is receiving from outside, inducing particular transcription factors to be active or repressed. If we could deconstruct development into each of the decisions a cell makes at different times in development, each of these decision points will look a lot more like that pathwork.

Wagner: When I was trying to get to grips with modularity I read a paper, which I can't now locate, published by a Belgian group who looked at a model of metabolic networks. In principle, we can represent all possible network structures. But under what conditions can I separate out one part of the network from the others so that it becomes a separate module? Their mathematical analysis showed that the only way you can do this is if the different parts of the network communicate through a bottleneck, exactly as you found with Svb. What you found here is an instance in terms of transcriptional regulatory networks of what they found with their metabolic model. The only architectural way of getting modules is to have part of the network separated with a bottleneck of very few inputs.

Wilkins: Is Svb a transcription factor?

Stern: Yes.

Wilkins: What does it regulate in either trichome development or oogenesis? Presumably it has a big role in oogenesis?

Stern: Yes. It is a complex locus in that it has multiple splice forms. Most splice forms include only the more 3' exons; they don't include the first exon, which is specific to the *svb* transcript. If you take a piece of the genome that just includes those 3' exons you can completely rescue the Ovo phenotype. All the regulatory regions and important functions for ovarian development are in that piece of DNA. We now know that *svb* regulates a diverse set of genes involved in actin reorganization and cuticle pigmentation to form the trichome (Chanut-Delalande et al 2006).

Wilkins: We have questioned the value of the term 'tinkering'. I'd like to question the value of the term 'macroevolution', at least in the way that you have used it. You have described a species-level difference that is also represented as an intra-specific genetic difference. If those morphological differences can either be within species or between species, is the term macroevolution used in this way useful? Svb exists as a genetic variant within *Drosophila*, but it also exists as this intraspe-cific difference between *D. sechellia* and the other species. Can a morphological difference be classified as a macroevolutionary one, versus one that is just a strain difference in *Drosophila*?

Stern: Let me clarify what that strain of *svb-ovo* is. It is an induced mutation that eliminates all of those exons. This is not a natural variant in any sense. The second thing is that we have looked for variation in trichome patterning within species and we don't find it. I don't see what the problem is that you have with macroevolution.

Wilkins: Macroevolution usually refers to big differences. It is often used to mean the big differences between taxa.

Stern: When I first saw this pattern, I thought it was huge: every segment of the animal had changed. When we mapped it we thought we'd get multiple small-effect genes. We were very surprised when it was a single gene. I don't know a better way of defining macroevolution. You have to draw the line somewhere. I showed the pathwork of *svb*. We know that many of the genes upstream will alter trichome patterns when they are mutated. We know that many of the genes downstream will alter trichome morphology. There are many ways to change trichome pattern and morphology, but in this evolved case it has been only by changes in *svb*.

Budd: How do you know that this is the case?

Stern: We mapped it.

Budd: How do you know that it arose as a variation in that particular gene?

Stern: What I know is that there were mutations that were fixed in *svb* that caused this phenotype. The original variance may have been created by something else, but this is non-discoverable

Budd: But you are making a certain assumption here.

Stern: No, I am saying that I have found that these mutations are definitely fixed in the *D. sechellia* lineage.

Cheverud: You are ascribing their effects to interspecific differences. Would they have had the same effect when they were in a segregating population, or are the effects that you see due to other interacting genes?

Stern: The recombination assay clearly tells us that the effects are entirely mappable to these small regions.

Cheverud: The differences are but the effects in the original population that diverged could be different. Gene effects are not fixed properties.

Wagner: Then you wouldn't be able to reproduce the effects in the *D. melanogaster* background.

Stern: The mapping crosses are all hybrid crosses between species. They work in very messed up genetic backgrounds.

Weiss: It sounds like you are saying that macroevolution can exist between species but never within species.

Stern: Of course not! It was either fixed by selection or it was fixed by drift. There are no alternatives. If it was fixed by drift it was in the population for a long time, and for some reason only mutations at *svb* that alter this trichome pattern are neutral. Or it was fixed by selection. Or it could be a mix. The mutations were either in the population briefly, because they were swept quickly through, or they were in the population a long time through drift. But they were in the population for some subset of the time after they diverged from a common ancestor. We know that there are a lot of genes that can alter trichome pattern. Presumably there are a lot of these genes with mutations at very low frequency in this population that can alter trichome patterns. When we do selection experiments on almost any trait in the lab, we almost always get a response. This is a little odd. But when we start mapping those traits we tend to find they have rather bizarre effects. Often they can have pleiotropic effects, or epistatic effects. We must remember what these alleles initially were. They were initially taken from a small subset and were probably selected from a low frequency. Selectively advantageous alleles are never at low frequency in populations. Things at low frequency are either neutral or deleterious, and this is what we are selecting on in artificial selection experiments.

Wilkins: Some advantageous mutations are initially at low frequency.

Wagner: Most of them never make it out of the low frequency band.

Stern: There is a mish-mash of various mutations and weird kinds of effects. This is what we select on, and this is what we say is intraspecific variation. I would argue that most of that variation doesn't look like the type of variation that gets fixed between species. Most of the time most of the differences between species tend to be caused by alleles with relatively additive effects, with low epistasis.

Cheverud: I disagree: when we do studies we always find epistasis.

Stern: You are working within species. That is my point. If you took different mouse species and mapped traits, I would argue that you wouldn't find as much epistasis, and perhaps even fewer loci.

Cheverud: I see a connection between within species and between species variation.

Stern: Where is your evidence for the same between species variance?

Cheverud: I think of it as the same variation.

Brakefield: Unfortunately, this discussion reflects the extent of our ignorance rather than of our knowledge. Revealing the extent to which the genetic variances within species inform us about differences between species is an exciting challenge for the future. The relevance of our own artificial selection experiments in a single stock of a butterfly species to explaining differences among species of the same genus is likely to depend on the extent to which variation segregating somewhere within natural populations of one, albeit widely distributed, species can provide the bases for differences observed among related species. Species typically exhibit substantial spatial and temporal variability among different local environments, and they often have leaky genetic bounds (for example, in *Heliconius* butterflies). Some, even perhaps many, of the alleles that differentiate related species may prove to be 'private' to one particular species at any one time. How frequently a specific allele underlying a difference in morphology between two related species is (completely) absent throughout one of the two species, and fixed in the other, is to me unclear; we need more studies, including of the type you are doing, but using more systems and different types of traits to clarify these issues. A high degree of privacy for key alleles may in time be found to apply more to traits connected directly to speciation and reproductive isolation, and less to ones more closely associated with adaptive responses to differences in environment. Thus, standing genetic variation within (widely distributed) species for phenotypic variation in traits characteristic of such adaptive responses, such as eyespot patterns in *Bicyclus* butterflies, may be more likely to include (many) copies of alleles that underlie differences in form among related species. I think that the answers to such issues are likely to depend on the specific traits, ranges in environment, and species concerned. In particular, it will also depend on the extent to which the traits considered—and the alleles concerned—exhibit pleiotropic interactions with other traits, including those involved in the evolution of reproductive isolation. However, in my view developing any generalizations about these issues must await more data.

Budd: D. sechellia is a rather interesting and unusual species. What is going on with the rest of the organism, with the underwiring of the trichomes, for example?

Stern: The muscles and nerves aren't attached to trichomes. They are cuticular structures. This species is found on islands in the Seychelles and it feeds on a fruit that is toxic to most other *Drosophila*. It is adapted to this toxic fruit.

Budd: Is there any phenotypic covariation across the trichomes?

Stern: Not that we have found. The whole life history is different.

Duboule: I want to ask you the same naïve question I asked Paul Brakefield. You have no idea what these trichomes may be useful for, and some species have them while others don't. In the Seychelle islands, is there a special species of birds that recognize trichomes? You mentioned that they may have evolved particular reproductive strategies, which in a way would affect the *ovo* gene. One might think of a system where touching up on this *ovo* gene would bring about trichomes, which would then be completely useless.

Stern: I can tell compelling ecological stories for which I have no data, but the real question is whether there is pleiotropy for selection on the reproductive system. We do have data on this because we have mapped the genetic events: they have half as many ovarioles as sibling species. There is nothing mapping near *ovo*, for ovariole number (Orgogozo et al 2006).

References

Chanut-Delalande H, Fernandes I, Roch F, Payre F, Plaza S 2006 Shavenbaby couples patterning to epidermal cell shape control. PLoS Biol 4:e290

Orgogozo V, Broman KW, Stern DL 2006 High-resolution quantitative trait locus mapping reveals sign epistasis controlling ovariole number between two Drosophila species. Genetics 173:197–205

The economy of tinkering mammalian teeth

Jukka Jernvall and Isaac Salazar-Ciudad

Developmental Biology Program, Institute of Biotechnology, Viikki Biocenter, P.O. Box 56, University of Helsinki, FIN-00014, Helsinki, Finland

Abstract. A central aim of evolutionary developmental research is to decipher the relative roles of ecological and molecular interactions in explaining biological diversity. Tetrapod teeth show diverse evolutionary patterns with a repeated increase in dental complexity, especially in response to herbivorous habits. Most extensively in mammals, dentition increases in complexity by elaborating morphology of individual teeth rather than increasing the number of teeth. Even though evolution of mammalian dentition is governed by ecology, recent evidence on molecular signalling suggests that many details and even some general evolutionary tendencies may be instigated by development. Specifically, iterative use of the same developmental modules, the enamel knots, may have facilitated developmentally efficient, or economical, elaboration of tooth shapes without substantially compromising the existing morphology. These kinds of developmentally influenced tendencies may be hypothesized to be typical to many organs and systems showing repeated evolutionary patterns.

2007 Tinkering: the microevolution of development. Wiley, Chichester (Novartis Foundation Symposium 284) p 207–224

Development of a multicellular organism is a dramatic process where a single cell gives rise to an integrated group of cells, tissues and organs. The most observable aspects of the emergence of the adult phenotype manifest the continuity of phenotype. Compared to this overall unfolding of the parental phenotype during ontogeny, evolutionary alterations of development are much more subtle. Yet, these variations on the parent theme are what, under natural selection, have been required for the evolution of biological diversity. Consequently, the challenge for detecting evolutionary tinkering of development is to identify how small changes are embedded within a process of much larger transformations.

Vertebrate dentition is eminently suitable for detecting evolutionary tinkering of development. Individual teeth are mineralized prior to eruption and function, thus aspects of tooth shape preserve information about development. On the other hand, the final tooth shape reflects the function and diet of an animal. Tooth shapes show the highest morphological diversity in mammals (Fig. 1). Particularly

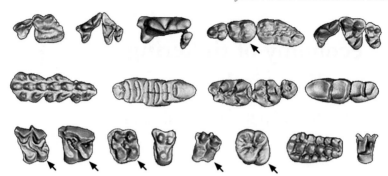

FIG. 1.   Examples of the diversity of mammalian teeth. Top, carnivoran cheek tooth rows; middle, rodent cheek tooth rows; and bottom, individual ungulate, human (third from the right), multituberculate and bat cheek teeth. Many of the different tooth types have evolved convergently, including the cusp in the distolingual corner of upper molars, the hypocone (arrows). Not to the same scale.

in mammals, commonly eaten foods can be inferred even from the details of a single tooth. In addition, by being highly mineralized, the fossil record of teeth is relatively complete. Perhaps not surprisingly, the functional principles of teeth have been recognized at least since Aristotle. At the dawn of modern paleontology, Cuvier (1825) argued that, by understanding the 'laws of the organic economy', the properties of a tooth can be used to explain the properties of many parts of the animal.

Whereas functional considerations have been recognized to play a significant role in the evolution of dental diversity, the exact nature of the available variation for natural selection to act upon, or the role of development, has remained more elusive. Bateson (1894) and several others (Butler 1952, Gould & Garwood 1969, Van Valen 1970) have observed regularities in the way tooth shapes vary along the tooth row and among species. These phenotypic regularities and gradients that allow inferences about the whole from its parts are suggestive about possible 'economy of developmental processes'. With a growing body of evidence on the molecular nature of patterning processes (Wilkins 2001, Gilbert 2006), it is now possible to examine the developmental basis of the phenotypic regularities. Accordingly, here we will not focus on the molecular mechanisms regulating tooth shapes *per se* (see Thesleff et al 2007, this volume), but rather how tooth development is modified during development and what kind of role development may have had on the evolutionary diversity of teeth.

Iterative subzoning of teeth

The basic morphological units of mammalian teeth are small prominences called cusps. The number, shape and location of cusps are especially diverse in molar

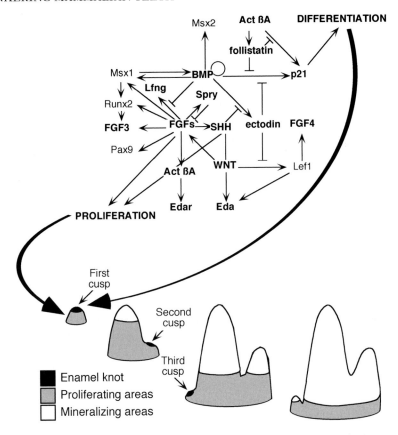

FIG. 2.   Partial signalling molecule network regulating cell differentiation and proliferation during tooth development (top). Epithelial cells differentiate initially locally and form signalling centres, the enamel knots which become cusp tips as the surrounding areas continue to grow (bottom). The enamel knot of the tallest cusp appears first followed by progressively shorter cusps. Before tooth eruption, epithelial cells will form the enamel whereas the underlying mesenchyme will form the dentin, resulting in a mineralized tooth that can change shape only due wear.

teeth (Hillson 2005, Fig. 1). During development, cusps appear at the locations where epithelial cells cease to divide and form epithelial signalling centres, the enamel knots (Jernvall et al 2000). Concomitantly, cells of the surrounding epithelium and underlying mesenchyme continue to divide, resulting in a folding and down-growth of the epithelium. One consequence of the folding process is the classic observations (Butler 1956, Berkovitz 1967) that in a tooth in which cusps differ in their height, the tallest cusps appear first during ontogeny followed by the progressively shorter cusps (Fig. 2).

The exact size and timing of the appearance of the enamel knots and cusps are now know to require a delicate balance of molecular signals activating and inhibiting enamel knot induction (Fig. 2). For example, mouse mutants lacking functional *ectodin* gene, an inhibitor of bone morphogenetic protein (BMP) signalling (Fig. 2), have enlarged enamel knots and the fully formed teeth have larger cusps and additional crests that join cusps together (Kassai et al 2005). Additionally, however, these *ectodin* mutants frequently have fused molars. This fusion is linked to the fact that *ectodin* is not only expressed around the enamel knots but also around the individual teeth (Kassai et al 2005). This experimentally-induced pleiotropic fusion of cusp and teeth underscores the iterative nature of tooth development. The available experimental and comparative evidence suggests that the secondary enamel knots, forming at the places of individual cusps, all express the same set of signalling molecules, and potentially the same set of transcription factors (Jernvall et al 2000, Fig. 3A). Furthermore, the secondary enamel knots appear to be replays of the preceding primary enamel knots that establish the crown area giving rise to individual teeth. The primary enamel knots, in their turn, are replays of even earlier signalling centres present during the establishment of the areas giving rise to the dental placodes, or regions forming teeth (Fig. 3A).

The iterative use of the same set of molecular signalling, or developmental modules, is best developed in mammalian molar teeth. During the past 55 million years, several mammalian groups have specialized in eating plants facilitated by an increase in the number of cusps on their molars. This increase in elaboration of tooth shape is often accompanied with decrease in tooth number, as is the case in elephants and rodents. The tendency for serially homologous structures to increase in specialization while decreasing in number, known as Williston's rule, has been identified in plants, insects, and vertebrates (Minelli 2003). In the case of dentition, the evolutionary elaboration results from iterative activation of the same developmental module, the enamel knots, within the existing spatial domain. This tendency can be called iterative subzoning where the same developmental module is repeatedly activated resulting in an iterative increase in complexity (Fig. 3B). In tetrapods through the Mesozoic, a repeated evolution of herbivory has been inferred to have occurred with characteristic increase in dental complexity. Whereas in the most Mesozoic forms (for example, in rhyncosaurs and cynodonts) increased chewing efficiency was accomplished by increasing individual tooth number, later forms (tritylodonts, multituberculates, eutherian mammals) increase cusp numbers within individual teeth (Fig. 3B).

Another tendency, perhaps causally linked to the iterative subzonation during development, is the convergent and parallel evolution of similar tooth morphologies among the Cenozoic mammals. Most notably, the squaring off of the upper molars was produced by the acquisition of the hypocone, the fourth cusp added to the primitively triangular tribosphenic molar (Fig. 1). The hypocone has

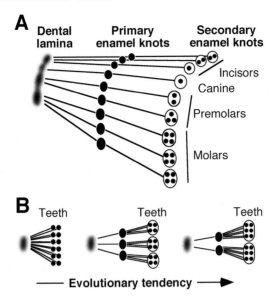

FIG. 3. In mammals largely the same signalling network (see Fig. 2) is redeployed from the tooth initiation to the formation of cusps. (A) In each iteration, the dental region is subdivided into new compartments and progressively larger numbers of the signalling domains are induced resulting in large numbers of enamel knots forming in complex cheek teeth. (B) In many tetrapod lineages evolution of plant-eating has been associated with increasing dental complexity. Whereas initially chewing efficiency was accomplished by increasing individual tooth number (on the left), later forms increase cusp numbers within individual teeth by iterative subzoning of the existing spatial domain during development.

appeared at least 20 times in different mammalian lineages, and its evolution has been reconstructed to signify the onset of a shift from animal dominated diet to the consumption of plants (Hunter & Jernvall 1995). The details about how the hypocone evolves can differ; most frequently the hypocone emerges from the posterior ridge at the base of the tooth, or as an enlargement and posterior–distal shift of another small cusp, the metaconule. Despite the differences in the way that the hypocone can evolve, the resulting morphologies can be so similar that the mode of hypocone origin can not be inferred from the morphology of the tooth alone.

Economy of developmental tinkering

Together the iterative subzoning of dentitions and the repeated evolution of specific morphologies could indicate a form of developmental 'economy', in the sense of frugality, in the regulation of tooth shape. This also suggests that at the level of microevolution, the evolutionary elaboration of tooth shapes could be understood if we can explain how new cusps appear in populations. Whereas it is fairly safe to state that there is a general agreement that the evolution of multicusped teeth is linked to an increase in functional integration of opposing teeth, early stages of multicusped teeth has not involved refined occlusion (e.g. Kielan-Jaworowska et al 2004). Indeed, in many living mammals small accessory cusps reach occlusal contact only when the jaw is almost completely closed and some short cusps may not even show wear until large cusps are substantially worn (Fig. 4A). Furthermore, short cusps are typically highly polymorphic in populations, suggesting indirectly that perhaps one reason for the evolution of multicusped teeth has been simply that there have not been reasons against it.

Because short cusps are developmentally the last ones to form during tooth crown formation, the iterative nature of cusp development is most easily appreciated in teeth that have cusps that are of different height. For example, postcanine

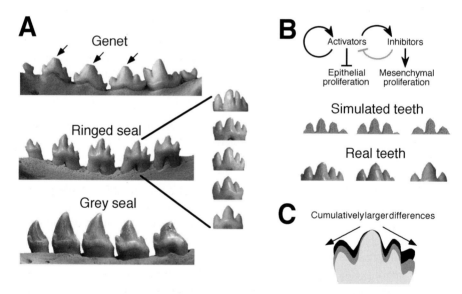

FIG. 4.   Tinkering with iterative formation of cusps. (A) In simple mammalian teeth, such as in the premolars of many carnivorans, there are typically short accessory cusps anteriorly and posteriorly to a tall central cusp. The number and size of the short accessory cusps is highly variable among species and also within species, as shown for the third postcanine tooth of the ringed seal on the right. Note that in the dentition of genet (a viverrid), the tall premolar cusp tips are worn flat (arrows) whereas the short cusps are still unworn. (B) A mathematical model simulating tooth development shows that by adjusting a single parameter in the activator-inhibitor loop (grey connection in the model), variation comparable to real seal teeth can be produced. (C) A patterning cascade mode of tooth formation requires only minimal developmental tinkering and can substantially alter short cusps while barely affecting tall cusps.

dentition of many carnivores, including seals, has one tall central cusp with progressively shorter cusps anteriorly and posteriorly (Fig. 4A). When individual cusps are replays of the enamel knots, a developmental patterning cascade results in which variation is cumulative (Jernvall 2000). That is, the position, size and number of the late-forming short cusps depends on the position, size and number of early-forming tall cusps. The effect of patterning cascade is also evident in mouse mutants where the last developing crown features are most affected (Grunberg 1965, Kangas et al 2004).

The repeated use of the same developmental module, the enamel knot, during development suggests that 'limited instruction sets' regulating the patterning cascade could be used to account for the appearance of new cusps. This possibility has been explored using mathematical modelling and by testing how simple gene networks can simulate the documented process of real tooth development (Salazar-Ciudad & Jernvall 2002). This kind of modelling approach is not focused on the importance of individual genes *per se*, but on the dynamics of the gene networks regulating tooth development. For example, modelling can be used to test whether similar phenotypic effects can be produced with alternative genetic pathways.

Analyses of the simulated teeth using the model indicate that tinkering with the parameters of a minimal activator–inhibitor loop, coupled with regulation of cell proliferation, easily produces an increase in cusp number (Salazar-Ciudad & Jernvall 2002, 2005). Furthermore, a single model parameter, corresponding to an inhibitor of enamel knot induction, can produce variation comparable to the variation found in real teeth (Salazar-Ciudad & Jernvall 2002, Fig. 4B). A large increase in the number and height of the short cusps will be possible with small changes in the spacing of all the enamel knots, thus barely affecting the positions of the tall, first developing cusps (Fig. 4B,C). This kind of change in the patterning cascade requires only small tinkering of the gene circuits regulating development. At the same time, the variation produced has minimal effects on the features that are likely to be most critical for the survival. Taken together, it could indeed be postulated that there are very few reasons against, and quite a few for, variation that produces new cusps. When we consider the overall patterns of increased elaboration of molar shapes during evolution, development may have predisposed tetrapod dentitions to iterative subzoning and progressively more complex teeth. Furthermore, the relative lack of variation in the tall cusps could easily be interpreted as lack of pleiotropy, underscoring the benefits of including developmental biology considerations when studying phenotypic variation.

Obviously the variability of small cusps does not mean that large cusps remain static through time. Polly and co-workers have recently shown that tooth shapes, including position of large cusps, can diverge relatively rapidly among populations as long as the occlusal fit is maintained (Polly 2001, Polly et al 2005). In addition to suggesting that there might be plenty of available variation to allow small cusps to appear repeatedly over microevolutionary time, Polly's work indicates that what might be selected for under natural selection is occlusal fit rather than tooth shape *per se*. Indeed, whereas the number of studies addressing correlation between occluding teeth remain rare, both developmental (Marshall & Butler 1966) and phenotypic (Kurtén 1953) evidence shows that occluding parts of the dentition correlate perhaps more closely than adjacent parts within mandible or maxilla.

In addition to exploring variation of individual teeth, computer simulations allow the generation of large numbers (in the order of millions) of simulated teeth. Such an approach has been used to explore how model parameter space maps onto the dental morphospace (Salazar-Ciudad & Jernvall 2004, 2005). Whereas the obviously circumstantial relationship between the parameter space and the morphospace allows inferences about nature of the relationship between the genotype and the phenotype. For example, 'mutations' in model parameters produce progressively larger jumps in the final tooth morphology when the 'wild type' tooth shape is complex (Salazar-Ciudad & Jernvall 2005). On the other hand, mutation of simple teeth in the model produce more gradual variation. If real tooth development would obey a similar relationship between the genotype and the phenotype, evolution of progressively more complex, multicusped teeth could result in a complexity trap where no further changes in tooth shape are likely to appear without substantial fitness cost (Salazar-Ciudad & Jernvall 2005). This, together with the progressive refinement of morphology due to iterative subzoning, suggests that the same developmental system that may have facilitated evolution of new cusps could eventually also stifle further change of the most complex morphologies. While the complexity trap is not experimentally or empirically proven, these possibilities may bear some light to the discussion about the appearance of orders, or even phyla, early in evolutionary radiations (for example, see discussion in Gould 1989).

Conclusions

Current evidence suggests that evolutionary increase in dental complexity may have been facilitated by development. First, iterative use of the same developmental module, the enamel knots, may have facilitated economically efficient elaboration of tooth shapes. An effect of this developmentally driven tendency has been iterative subzoning of teeth where patterning happens progressively within the previous spatial domain. Second, a sequential patterning cascade of enamel knots allows addition of new cusps without substantial changes in the existing tooth morphology. Taken together, even though evolution of dentitions can be understood to be governed by ecology, details and actual morphological solutions may also be governed by the economy of developmental tinkering.

References

Bateson W 1894 Materials for the study of variation, treated with special regard to discontinuity in the origin of species. Macmillan, London

Berkovitz BKB 1967 The dentition of a 25-day pouch-young specimen of *Didelphis virginiana* (Didelphidae: Marsupialia). Arch Oral Biol 12:1211–1212

Butler PM 1952 Molarization of the premolars in the Perissodactyla. Proc Zool Soc Lond 121:819–843

Butler PM 1956 The ontogeny of molar pattern. Biol Rev 31:30–70

Cuvier G 1825 Discours sur les révolutions de la surface du globe, et sur les changemens qu'elles ont produits dans le règne animal. Dufour et d'Ocagne, Paris

Gilbert SF 2006 Developmental biology, 8th edition. Sinauer Associates, Sunderland

Gould SJ 1989 Wonderful life: the Burgess Shale and the nature of history. Norton, New York

Gould SJ, Garwood RA 1969 Levels of integration in mammalian dentitions: an analysis of correlations in *Nesophontes micrus* (Insectivora) and *Oryzomys couesi* (Rodentia). Evolution 23:276–300

Grunberg H 1965 Genes and genotypes affecting the teeth of the mouse. J Embryol Exp Morphol 14:137–159

Hillson S 2005 Teeth. Cambridge University Press, Cambridge

Hunter JP, Jernvall J 1995 The hypocone as a key innovation in mammalian evolution. Proc Natl Acad Sci USA 92:10718–10722

Jernvall J 2000 Linking development with generation of novelty in mammalian teeth. Proc Natl Acad Sci USA 97:2641–2645

Jernvall J, Keränen SVE, Thesleff I 2000 Evolutionary modification of development in mammalian teeth: Quantifying gene expression patterns and topography. Proc Natl Acad Sci USA 97:14444–14448

Kangas AT, Evans AR, Thesleff I et al 2004 Nonindependence of mammalian dental characters. Nature 432:211–214

Kassai Y, Munne P, Hotta et al 2005 Regulation of mammalian tooth cusp patterning by ectodin. Science 309:2067–2070

Kielan-Jaworowska Z, Cifelli RL, Luo Z-X 2004 Mammals from the Age of Dinosaurs: Origins, Evolution, and Structure. Columbia University Press, New York

Kurtén B 1953 On the variation and population dynamics of fossil and recent mammal populations. Acta Zool Fennica 76:1–122

Marshall PM, Butler PM 1966 Molar cusp development in the bat, Hipposideros beatus, with reference to the ontogenetic basis of occlusion. Arch Oral Biol 11:949–965

Minelli A 2003 The development of animal form ontogeny. Morphology and evolution. Cambridge University Press, Cambridge

Polly PD 2001 On morphological clocks and paleophylogeography: towards a timescale for Sorex hybrid zones. Genetica 112:339–357

Polly PD, Le Comber SC, Burland TM 2005 On the occlusal fit of tribosphenic molars: Are we underestimating species diversity in the Mesozoic? J Mamm Evol 12:285–301

Salazar-Ciudad I, Jernvall J 2002 A gene network model accounting for development and evolution of mammalian teeth. Proc Natl Acad Sci USA 99:8116–8120

Salazar-Ciudad I, Jernvall J 2004 How different types of pattern formation mechanisms affect the evolution of form and development. Evol Dev 6:6–16

Salazar-Ciudad I, Jernvall J 2005 Graduality and innovation in the evolution of complex phenotypes: insights from development. J Exp Zool B Mol Dev Evol 304B:619–631

Thesleff I, Järvinen E, Suomalainen M 2007 Affecting tooth morphology and renewal by fine-tuning the signals mediating cell and tissue interactions. In: Tinkering: the microevolution of development. Wiley, Chichester (Novartis Found Symp 284) p 142–157
Van Valen L 1970 An analysis of developmental fields. Dev Biol 23:456–477
Wilkins AS 2001 The evolution of developmental pathways. Sinauer Associates, Sunderland

DISCUSSION

Bell: An essential feature that you left out is that there is the primary knot, and if it is large enough it divides into secondary knots. But exactly where these are within the field of the primary knot really matters. What determines this?

Jernvall: The modelling and the evidence we have show that it is a dynamic process. It could be that we can change the position of these knots by altering how strongly they see inhibitors, for example. There are also gradients of signalling molecules. It is a dynamic process with no predetermination of the position of the cusps.

Bell: Are there standing waves of intersections of different signals that determine this positioning?

Jernvall: It would be generated by the tooth itself. How close an enamel knot and cusp can form depends on inhibition, for example. Irma Thesleff showed earlier that activated Wnt signalling induces the formation of small enamel knots close to each other with more teeth forming in a small space.

Wagner: I buy this model, but at the macroevolutionary level it seems not that easy. Most reptiles make conical teeth, and it is primarily in the mammals that we see complexity emerging in the dentition. There must be something different between these two types of teeth, with the mammalian ones lot more tinkerable and one less so, in terms of cusps.

Jernvall: We know crocodiles do have enamel knots. At some level the development is similar. Then one gets into metabolics and explanations for why mammals have evolved complex teeth.

Wagner: Many reptiles are also carnivores, and they might want to chew their prey rather than swallow them whole.

Lieberman: Until unilateral mastication evolved there wasn't any selection advantage to develop this complex dentition.

Donoghue: Uromastyx, an agamid lizard is an example of a lizard that has attained a complex interpenetrative occlusal dentition independently of mammals.

Jernvall: Yes, this has happened in the past.

Carroll: I was going to go further back and look at amphibians. Most amphibians have very simple teeth, but some form tooth plates, with lots of cusp-bearing separate teeth. But there are a few amphibians that had three cusps like trichonodonts. There is a variable capacity to evolve this. Presumably, none of these amphibians are herbivores. The capacity to form a series of small cusps is there globally.

Weiss: There are some fish with multiple cusped teeth.

Bell: Cichlids can have incredibly complex patterns on the mandibular teeth (Fryer & Iles 1972, Barlow 2002).

Weiss: People have done reaction–diffusion modelling of reptilian teeth that only have a single cusp, and have got good fits between theory and the actual pattern. So there is some internally nested aspect to this.

Jernvall: There seems to be something more versatile about mammalian dentition. To make a modern mammalian tooth there has to be a base of the tooth crown. This could be something that has evolved later.

Wagner: If I think about it in microevolutionary terms, I would ask how much natural variation we find in living reptile populations for the number of cusps. Even if this variation might not be adaptive, does it occur regularly enough that if the selection pressure arises there could be a selection response?

Weiss: After multiple cusps evolved in mammals, it took another 150 million years before complicated teeth are seen. Why did it take so long?

Jernvall: It took 40 million years from when increased cusp number was first seen to get what we could call teeth with exact occlusion. There are two extreme answers. One is that something must have evolved in development to allow the evolution of complicated teeth. The other answer is ecological: the early mammals may have had no selection towards complex teeth.

Lieberman: There's a third answer: the cusps could be part of a larger system— the occlusal system. There could have been all kinds of modifications that first had to evolve in the origins of the jaw joint and the ability to have unilateral mastication; these then allowed selection to act on the variation that existed. You mentioned that adding more cusps can be done without compromising fitness. In thinking about occlusion, I can see that within one tooth this would be the case, but teeth have to fit together. The lower big cusp may fit in between two upper teeth, so adding cusps affects how other teeth then occlude.

Jernvall: You can add little cusps as long as they don't start to affect occlusion. There is always room for a little bit of modification.

Oxnard: My understanding is that in seals the milk dentition is shed in the caul (birth membranes) before birth. They are born with the permanent dentition in place. Does anyone know what those shed milk teeth are like?

Jernvall: Yes, seals are an example of animals with vestigial milk dentition. The milk teeth of seals are reduced to a simple cone with little sign of small cusps.

Hanken: In the early 1980s when the idea of developmental constraints was popular, there were a series of studies that, for a particular developmental system and model of development, would assess naturally occurring variation and classify variants into one or two groups: 'forbidden' versus 'permitted' morphologies (e.g. Stock & Bryant 1981, Holder 1983). In your case you have shown that many of the cusp patterns are permitted. Do you know of any cusp patterns in mammalian dentition that seemingly are forbidden but nevertheless exist? Such examples could be the exceptions that prove the rule.

Jernvall: This is what we try to look at in seals. Our prediction is that it would be very difficult to find a tooth where there is a tall cusp and then the second tallest cusp is missing and a small third cusp is present, or conversely, you would have a disproportionally small third cusp next to a tall second cusp. I am still looking for examples of these in seals. I am sure they have them, but they must be very rare. That is, some variation is much more frequent than other types.

Wilkins: Didn't you publish a paper several years ago talking about the much greater variety of mammalian teeth that used to exist (Jernvall et al 1996)?

Jernvall: Yes, we looked trends in the evolution of ungulate cheek tooth shapes during the Cenozoic.

Wilkins: This increases the sample size of possible tooth types. Yet even with this increased sample size you haven't found certain morphologies, suggesting that they are 'forbidden'.

Jernvall: At that point we didn't understand enough about development for us to formulate predictions about what tooth shapes would be forbidden.

Coates: You talked a lot about crowns. What is known about patterning of roots? The root systems seem to be a lot more conservative. In the palaeontological community there are several specialists who study fossil shark teeth, and erect large numbers of taxa on the basis of crowns. But they do seem to be a little more conservative when they start characterizing tooth bases, suggesting that variation of the (tooth) root system is more constrained.

Jernvall: To a degree, the position and number of roots seem to follow the number of cusps. It almost looks opportunistic, that is, roots form to fill in the space formed previously during the crown development.

Lieberman: It is important in primates. People use differences in the number of roots to establish species. This is a *post hoc* use of morphology without understanding its development.

Bell: It is important for you, but it may not be for the animals.

Wagner: Irma Thesloff's paper showed us that the tooth types, premolars and molars, are derived from one tooth germ that then sprouts. This is another layer of connections that you didn't show in your schemes.

Jernvall: It happens for molars but it doesn't look like this is the case with premolars. Premolars form from a continuous lamina whereas molars are posterior extensions where a posterior molar 'sprouts' from the previous molar.

Weiss: There is a regional zonal difference underlying this. A gene, *Barx1*, seems to be necessary for multicusp teeth.

Ackermann: I am coming from a human evolution background where we tend to think about teeth as being relatively constrained developmentally. Whenever there are differences in cusp morphology it tends to be attributed to something meaningful, such as adaptive divergence. So it just makes me wonder how variable a tooth can be *within* a population—how much 'play' is there, if you will—and whether it is justifiable to attribute small differences in cusp morphology between individuals to macroevolutionary adaptation or rather underlying developmental variation within a species.

Jernvall: The first step to get to this is to sample variation in different kinds of teeth. The way seals use their dentition is for grabbing fish or invertebrates. One would imagine that if selection is relaxed, which is likely to be the case in seals, we might get more developmental variation and might see more of the tinkering in action.

Brakefield: Has anyone looked at the differentiation of tooth morphology in different breeds of dogs.

Jernvall: We need to look at this. The problem with seals is that we can't do developmental biology on them. The obvious extension of this study is dog premolars which are relatively close to seals both morphologically and phylogenetically.

Hall: All of your studies have been based on the lower jaw. Have you looked at correlated variation in the occluding teeth of the upper jaw?

Jernvall: To some extent. Whenever the cusps become more equal in height in the lower teeth, the same thing happens in the upper teeth. There seems to be a strong correlation.

Hall: I can imagine some hybridization studies between different strains that might be interesting where the patterns are different. You might get some interesting variation popping up. I have another question. Is there a change in the mode of tooth attachment between reptiles and mammals? It might be interesting to look at this transition. It might affect the rate at which cusps were modified.

Cheverud: Do your simulations and models make any predictions about the biases in levels and directions of variations in the teeth. Are some variations more likely than others?

Jernvall: Yes, and this seems to be in part due to the morphodynamic nature of the model. As soon as you couple the molecular signalling and growth, the system becomes much more non-linear so it is more difficult to predict the outcome. For example, if you change one parameter gradually, nothing might happen for a while, and then there is a jump from one shape to another. If you think about a change between parallel and diagonal cusp positions, there is some indication from the model that this shift is likely to be a rapid jump with few intermediates.

Wilkins: I have a question about the early stages of tooth evolution. The tooth formation mechanism is related to the mechanism that forms scales and feathers. Has there been a comparison with the networks involved in tooth formation in fish with scale formation in fish? What were the first steps in developing teeth specifically as opposed to scales?

Donoghue: There is some information becoming available. All that we know is at the level of generalities of the cell-signalling molecules and transcription factors that are involved in the patterning of epithelial appendages.

Hall: The only study that I know is one by Gareth Fraser comparing molecules involved in fish teeth versus mammalian teeth (Fraser et al 2006).

Wilkins: Are there any data on how the networks differ yet?

Hall: No, this was just looking at a couple of major conserved upstream molecules such as Wnt and BMP.

Thesleff: One special feature of the teeth is that the mesenchyme has to be from neural crest. This seems not to be true for the other appendages.

Donoghue: It is assumed. The only experiments have been on lepidotrichia, which are more similar to scales.

Carroll: At the base of fish, there are no teeth. It used to be suggested that teeth had evolved from scales moving over the jaw margin. Now there is evidence that these may have been pharyngeal teeth from the mouth lining working their way out.

Donoghue: The evidence for this is weak. We know primitively that fish have scales but they don't have teeth. Some derived fish have pharyngeal scales. Going back to the neural crest issue, we know that primitively there is dentine in dermal scales and what we know of dentine development is that it is contingent on neural crest.

Coates: The earliest mineralized tissues found in a chordate are within the oral cavity. These are the conodonts, the tooth-like arrays found in the mouth of the conodont animals.

Weiss: People fight vigorously over whether this counts or whether it is independent and separate.

Hall: Hair and teeth might be a simpler comparison to make, as opposed to fish scales and teeth. We know a lot about the signalling systems in hair development, and there is a lot of conservation there with teeth. I don't know about the level of the networks.

Wilkins: All of these structures are in some sense novelties, yet they are clearly related. It would be nice to begin to put some genetic basis to what makes particular novelties, if one can.

Hall: This gets the tinkering notion to the level of macroevolution.

Coates: I was interested in Williston's rule—the idea that serial homologues start off with a set of multiple and general items, and end up with a reduced number that are individually specialized. I feel cautious about this, because it is loaded with the idea that undifferentiated equals primitive, and that tinkering must start on something very general in form. It's a developmentalist's view of evolution that regards an undifferentiated state as a primitive condition.

Hall: I haven't seen Williston's rule written in quite the way you have presented it. I think of it in terms more of the number of elements.

Morriss-Kay: This is related to one of the questions posed at the beginning of the meeting: how true is von Baer's law at the microevolutionary level?

Hall: This is why I was asking Rudy Raff yesterday about how early he could chase back the changes in the coelom. If we could chase it back to lineages and things that happened very early this might indicate von Baer was out to lunch with his conception of early development being highly conservative.

R Raff: As far as Williston goes, consider the snake. It goes from fewer segments to more and from more complicated ones to less complicated ones. It doesn't fit the rule.

Hall: It seems clear from what you described that you can get major changes early in development. There are changes in cell lineage very early in development controlled by the maternal genome.

R Raff: This was ignored for a long time because the concepts that were coming from the Haeckelian period were that early development is not something that is going to evolve much. Phylogenetic charts from that period like to show a single cell which gives rise to the blastacean, gastracean and diplural larvae, but these larvae and embryonic forms are seen as ancestral animals. Then, suddenly you start seeing as phylum-level adult distinction.

Hall: A lot of these very early changes, for example in cell lineage, are the maternal factors: it is really the maternal genome that is making changes to redirect cell lineages. Then the zygotic genome is doing things later on. There are two different genomes here when we are talking about tinkering. This adds another level of complexity.

R Raff: There is a phenomenon seen in molluscs and annelids: equal and unequal cleaving. Unequal cleavers produce a polar lobe. This is maternally determined, so that the two cells that result have different developmental potentials as a result of absorption of the polar lobe by one of them. In the equal cleaving annelids and molluscs this doesn't happen at all, and the other quadrants of the embryo are set up by cell–cell interactions at the four cell stage. There don't seem to be constraints: this seems to go back and forth between maternal and zygotic determination among these groups. It sounds like a difficult mechanism to switch back and forth, but it doesn't seem to be. It is a phenomenon of extreme flexibility in one of the most important determinant events of very early development.

Budd: Despite what you say, it would be idle to pretend that changes in early developmental events are not usually rather bad. What, empirically, do these instances of early flexibility tell us?

Lieberman: That's why I wrote the question 'how often?' in my introduction. When von Baer's law is discussed people throw out examples that refute it. Has anyone done a quantitative study of it?

Cheverud: About two-thirds of death occurs before implantation. Most is to do with chromosomal abnormalities.

Lieberman: This is a point raised by Jacob in his essay.

R Raff: One group of animals we are fond of, the amniotes, arose by an extremely early developmental change.

Wagner: How well do we know the phylogeny and the developmental modes in these invertebrates? My experience is you need an incredibly good taxon sampling to make statements about rates of transitions. I just don't think we know enough.

R Raff: There is a bunch of related groups where this sort of thing has been demonstrated.

Wagner: With vertebrates you can mess around in early development and later development is quite robust, but then you come to the phylotypic stage where any messing around is less tolerated.

Weiss: If you look at a lot of mouse embryos, they vary tremendously. Staging mouse embryos is difficult because there's a lot of variation in viable embryos. Michael Richardson says that there is some truth to this phenotypic similarity at what has been called the phylotypic stage; in other words, despite exaggeration it was not totally made up by Haeckel.

Lieberman: There's not much measurement of variation at most developmental stages.

R Raff: In fact there is repression of it. There is the famous cell lineage of the frog, *Xenopus*. It turns out that in order to get the cell lineage the investigator had to throw away most of the embryos and choose the ones where the first cleavage went right through the grey crescent. Otherwise there is a different cell lineage from every other sampling.

E Raff: She said that in her paper; it is the text books that ignored it.

Hallgrimsson: There is a counter example to the argument that there is lots of variation early in development and most of it is lethal. In food animal species, for example cows, they will put cows and bulls together for one cycle and 98% of them get pregnant. Clearly, there is not much embryonic death in these species. Presumably either the magnitude of variation in early development or the suscep-tibility of later development to that variation has been acted on by artificial selec-tion in these species.

Wagner: In terms of measurement there is a paper by Galis & Metz (2001) which shows that the severity and frequency of toxic effects are much higher during the phylotypic stage than during other phases of development. There is a clear window, coincident with the phylotypic stage, where the effects are more frequent and more severe, than earlier or later in development.

Weiss: What about flies? How variable are the early stages?

Stern: There have been some dramatic experiments manipulating early pattern-ing and adding more copies of the *bicoid* gene (Namba et al 1997). The early pat-terning events can be altered dramatically and the fly emerges happily at the end; the whole system regulates itself.

Bell: You have described a kind of buffering. Why is that there? How did it evolve?

Stern: There is some nice computational work showing that the segmentation pathway is remarkably robust to alterations in parameters (von Dassow et al 2000).

Wagner: This was just a couple of paragraphs long. Then he wrote a longer paper showing that embedded in there is a small network that can produce the same outcome but is much more brittle. This looks to me like an evolved robustness.

Wilkins: There is experimental evidence using haploinsufficient strains that argues against this (Galis et al 2002). This has to do with making more than one segment polarity gene haploinsufficient. You start getting all sorts of problems and lethality.

Hall: There is a tremendous redundancy built into development. Why do we need half a dozen morphogens to induce mesoderm in frogs?

Wagner: There is a robustness-related add-on to what I discussed in my paper. There is a follow-up of the study by Sean Carroll (Galant & Carroll 2002). He tried to see what happens if you put the velvet worm gene in place of the native *Dro-sophila* gene. He got almost perfect flies, except that a particular bristle was bifurcated.

Stern: I was involved in that work. They actually knocked out the C-terminal region including the conserved QA repeat region. It wasn't the whole protein. There were a couple of subtle changes.

References

Barlow GW 2002 The cichlid fishes. Perseus, Cambridge, MA

Fraser GJ, Graham A, Smith MM 2006 Developmental and evolutionary origins of the verte-brate dentition: molecular controls for spatio-temporal organisation of tooth sites in ostei-chthyans. J Exp Zoolog B Mol Dev Evol 306:183–203

Fryer G, Iles TD 1972 The cichlid fishes of the Great Lakes of Africa. Oliver and Boyd, Edinburgh

Galant R, Carroll SB 2002 Evolution of a transcriptional repression domain in an insect Hox protein. Nature 415:910–913

Galis F, Metz JA 2001 Testing the vulnerability of the phylotypic stage: on modularity and evolutionary conservation. J Exp Zool 291:195–204

Galis F, van Dooren TJ, Metz JA 2002 Conservation of the segmented germband stage: robust-ness or pleiotropy? Trends Genet 18:504–509 (erratum: Trends Genet 18:658)

Holder N 1983 Developmental constraints and the evolution of vertebrate digit patterns. J Theor Biol 104:451–471

Jernvall J, Hunter JP, Fortelius M 1996 Molar tooth diversity, disparity and ecology in Cenozoic ungulate radiations. Science 274:1489–1492

Namba R, Pazdera TM, Cerrone RL, Minden JS 1997 Drosophila embryonic pattern repair: how embryos respond to bicoid dosage alteration. Development 124:1393–1403

Stock G, Bryant S 1981 Studies of digit regeneration and their implications for theories of development and evolution of vertebrate limbs. J Exp Zool 216:423–433

von Dassow G, Meir E, Munro EM, Odell GM 2000 The segment polarity network is a robust developmental module. Nature 406:188–192

Pelvic skeleton reduction and *Pitx1* expression in threespine stickleback populations

Michael A. Bell, Kaitlyn E. Ellis and Howard I. Sirotkin*

*Department of Ecology and Evolution, Stony Brook University, Stony Brook NY 11794-5245, USA, and *Department of Neurobiology and Behavior, Stony Brook University, Stony Brook, NY 11794-5230, USA*

Abstract. The pelvic skeleton of threespine stickleback fish contributes to defence against predatory vertebrates, but rare populations exhibit vestigial pelvic phenotypes. Low ionic strength water and absence of predatory fishes are associated with reduction of the pelvic skeleton, and lack of *Pitx1* expression in the pelvic region is evidently the genetic basis for pelvic reduction in several populations. Pelvic vestiges in most populations are larger on the left (left-biased), apparently because *Pitx2* is expressed only on that side. We used whole-mount *in situ* hybridization to study *Pitx1* expression in 19 populations of *Gasterosteus aculeatus* from lakes around Cook Inlet, Alaska, USA. As expected, specimens from six populations with full pelvic structures usually expressed *Pitx1* in the limb bud; those from eight populations with left-biased pelvic reduction usually did not express it. Specimens from one of three populations with right-biased or unbiased pelvic reduction sometimes expressed *Pitx1*. One of two populations in which the pelvic spines (but not the girdle) are usually absent often expressed *Pitx1*. In terms of Jacob's 1977 'tinkering' metaphor, *Pitx1* was the spare part with which natural selection usually tinkered for stickleback pelvic reduction, but it also tinkered with other genes that have smaller effects.

2007 Tinkering: the microevolution of development. Wiley, Chichester (Novartis Foundation Symposium 284) p 225–244

In the mid-20th century, population genetic theory created a synthesis between evolution and genetics, and abundant allozyme and DNA sequence data soon added empirical support for this theory. However, the excitement created by population genetics eclipsed research on the relationship between ontogeny and evolution (Gould 1977). Powerful molecular and computational tools now allow a broader evolutionary synthesis that incorporates the developmental bridge between genetics and morphology (e.g. Raff & Kaufman 1983, Stern 2000, Carroll et al 2005). While Jacob's (1977) notion of tinkering in evolution is implicit in

Darwinian selection and population genetics, de-emphasis of the developmental link between genotype and phenotype in the New Synthesis ignored the developmental mechanism by which variation for this tinkering is produced. Early contributions to the current synthesis of development and evolution have usually compared distantly related species because the genomic tools needed to compare closely related but phenotypically divergent populations were rarely available (see Stern 1998). However, studies of closely related populations are necessary to distinguish neutral genetic differences from those that caused morphological evolution and to investigate the developmental basis of Jacob's evolutionary tinkering.

The threespine stickleback (*Gasterosteus aculeatus*) is a small fish from Holarctic marine and freshwater habitats. Freshwater populations exhibit extraordinary morphological variation, and the biological properties of this species make it suitable for diverse studies, including development and genetics (Bell & Foster 1994). In the past decade, specialized methods and genomic resources have been developed to study stickleback developmental genetics (Peichel et al 2001, Kingsley et al 2004), justifying its characterization as a biological 'supermodel' (Gibson 2005).

Most threespine stickleback populations have a robust pelvic skeleton, which helps deter predatory vertebrates (e.g. Bell 1988, Bell & Foster 1994). However, *G. aculeatus* from isolated lakes with low ionic concentration and lacking predatory fishes may exhibit extreme reduction or loss of the pelvic skeleton (Bell 1987, 1988, Bell et al 1993; Figs. 1, 2). These populations are interspersed with populations with full pelvic structures throughout the northern portion of the species' range. They are restricted to recently deglaciated regions, and pelvic reduction never occurs in the marine stickleback from which these freshwater isolates have evolved (Bell 1987, 1988). Furthermore, pelvic reduction evolved in threespine stickleback from a 10 million year old fossil deposit (Bell et al 2006). Thus, pelvic reduction has evolved independently many times within freshwater populations. Nevertheless, the order in which components of the pelvic skeleton are lost is remarkably similar among modern populations (Bell 1987, 1988, Bell et al 1993).

The stickleback pelvic skeleton normally develops as endochondral bone in the lateral plate mesoderm and struts of dermal bone within the adjacent limb bud. Expression of *Pitx1* during pelvic skeletal development is critical for its formation. It is a homeodomain transcription factor that determines hindlimb identity (Marcil et al 2003). *Pitx1* knockouts in mice have hindlimbs that resemble a forelimb, and induction of *Pitx1* expression in the chick forelimb causes it to resemble the hindlimb (Szeto et al 1999, Logan & Tabin 1999). Genetic analyses in *G. aculeatus* from Scotland, Iceland, British Columbia, and Alaska implicate silencing of *Pitx1* expression in pelvic reduction of threespine stickleback (Cole et al 2003, Shapiro et al 2004, Cresko et al 2004).

Another homeobox transcription factor, *Pitx2*, apparently complements *Pitx1* regulation of pelvic development (Marcil et al 2003). In a variety of species, *Pitx1*

VENTRAL **LATERAL** **SCORE**

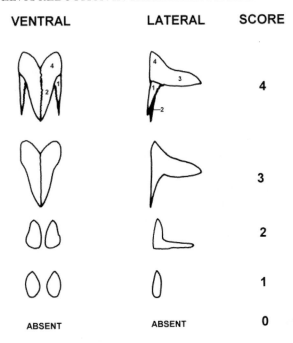

FIG. 1. Pelvic phenotypes and scores used to quantify pelvic reduction. Reproduced with permission from Bell et al, Evolution of pelvic reduction in threespine stickleback fish, Evolution 47:906–914. © 1993, Copyright Arizona State University.

knockouts develop asymmetrically reduced hindlimbs that tend to be larger on the left due apparently to expression of *Pitx2* on only that side (Meyers & Martin 1999, Marcil et al 2003). Pelvic vestiges in most threespine stickleback with pelvic reduction tend to be larger on the left, implicating *Pitx1* (Francis et al 1985, Cole et al 2003, Shapiro et al 2004, 2006a, Bell et al 2007).

Cook Inlet, Alaska, USA contains countless lakes with threespine stickleback. It was almost completely glaciated until about 22 000 years ago (Reger & Pinney 1996), and freshwater stickleback there must have evolved from marine ancestors since deglaciation (Bell et al 1993). Bell & Ortí (1994) reported about 40 Cook Inlet populations of threespine stickleback with pelvic reduction. Asymmetrical specimens usually have a larger left pelvic vestige (i.e. 'left-biased'), but those in four populations are more likely to have a larger right pelvic vestige (i.e. 'right-biased') or to exhibit fluctuating asymmetry (Francis et al 1985, Bell et al 2007). In addition, most specimens in some populations lack both pelvic spines but have the full pelvic girdle (Bell & Ortí 1994). We used whole-mount *in situ* hybridization to study the role of *Pitx1* expression in pelvic reduction in 19 Cook Inlet stickleback populations with diverse pelvic phenotypes.

L-Shaped   Morvoro

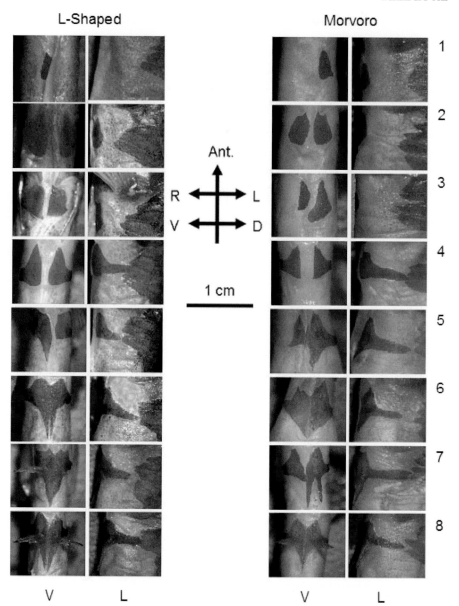

FIG. 2.   Morphological series for pelvic reduction in threespine stickleback from L-Shaped (right-biased) and Morvoro lakes (left-biased) showing the similarity of pelvic phenotypes in right and left-biased populations. V and L indicate ventral and lateral views. Anterior (Ant.) is up in each image, R and L indicate the direction of right and left in the ventral images, and V and D indicate ventral and dorsal in the lateral images.

Materials and methods

Selection of populations

Stickleback populations (Table 1) were selected on the basis of pelvic morphology (Bell & Ortí 1994) and directional asymmetry (Bell et al 2007). We used six populations with monomorphic full pelvic expression (pelvic score [PS] = 8; Fig. 1), two with only the pelvic spines missing ('spineless,' PS = 6), and 11 with extreme pelvic reduction (including PS = 0–2). Populations with extreme pelvic reduction included eight left-biased and two right-biased cases, and one had fluctuating asymmetry. The genetics of pelvic reduction was studied in three of the left-biased populations we used (i.e. Bear Paw, Boot, Whale) by Cresko et al (2004). Sample locations are in the Matanuska-Susitna Valley or on the Kenai Peninsula, Alaska, and were reported in Bell & Ortí (1994) and Bell et al (2007), except for Rabbit Slough, which is at 61.534 N Lat and 149.268 E Long.

Capture and preparation of adults and juveniles

Adult and juvenile stickleback were captured using minnow traps (Bell et al 1993). Specimens for morphological analysis were anaesthetized in MS-222, fixed in 4% dolomite-buffered formaldehyde solution, transferred to 50% isopropyl alcohol, and stained with alizarin red S to visualize the pelvis (Bell et al 1993). Mature adults for genetic crosses were transported to the laboratory live in ice chests containing aerated lake water and crossed within a few days.

Adult and juvenile pelvic phenotypes

Pelvic phenotypes can be scored when threespine stickleback reach 20 mm standard length (Bell et al 1993). We scored pelvic phenotypes from Zero Lake ($n = 155$) and Rabbit Slough ($n = 2144$) and pelvic asymmetry from Echo Lake fish ($n = 100$) by inspection (Table 1, Fig. 1). Other phenotypic information came from Bell & Ortí (1994) and Bell et al (2007). Digital images of pelvic phenotypes were captured and enhanced using Adobe Photoshop to darken features without changing their shape.

Selection of parental phenotypes for crosses

All specimens from populations without pelvic reduction had a full pelvic girdle and both spines (PS = 8). Parents from populations with only spine loss lacked both spines (PS = 6). Most parents from populations with extreme pelvic reduction had no or small pelvic vestiges (PS ≤ 2), but a minority had higher pelvic scores (Table 2).

TABLE 1 Pelvic girdle phenotypes (Bell and Ortí 1994) and strength of directionality among asymmetrical specimens (Bell et al 2007) of threespine stickleback populations used in this study

Population	Pelvic score frequency									Mean pelvic score	% left larger
	0	1	2	3	4	5	6	7	8		
Bear Paw	7.6	15.2	65.2	3.8	5.7	2.4				1.92	96
Beaverhouse									100.0	8.00	
Boot	8.4	9.2	66.4	5.2	8.2	1.1	1.1	0.3		2.05	61
Bruce	4.4	4.0	10.3	7.0	29.0	12.5	27.9	2.9	1.8	4.27	64
Coyote	32.7	25.2	40.2	1.1	0.6			0.3		1.13	84
Dolomite	0.4		72.2	2.0	10.4	3.5	11	0.4		2.78	50
Echo	1.5		0.3		0.6	0.3	82	5.9	9.2	6.27	56
Kidney						0.7	61	21	18.0	6.55	
L-shape		1.6	17.1	5.7	24.0	3.3	8.9	8.1	32.0	5.28	35
Loberg									100.0	8.00	
Long									100.0	8.00	
Lynda									100.0	8.00	
Morvro	4.8	8.1	14.5	1.6	13.0	4.8	8.1	4.8	40.0	5.23	91
Nancy									100.0	8.00	
Orphia			4.9	2.4	32.0		59		2.4	5.15	0
Rabbit Slough									100	8.00	
Wallace	5.6	7.2	30.8	4.8	14.0	3.6	3.6	3.6	26.8	4.18	80
Whale	39.9	29.2	29.2	0.9	0.9					0.94	82
Zero			16.1	3.2	9.7	7.0	27.7	3.9	32.3	5.54	

See original studies for sample sizes.

TABLE 2 *Pitx1* expression in the limb bud of wild-caught fry (W) and families (F) from crosses

Population type	Source	Sample type	Parental phenotype ♂	♀	Number stained	Sample size	Percent stained
Full	Beaver House	W			6	6	100
	Loberg	W			80	82	98
	Lynda	W			39	40	98
	Long	W			3	3	100
	Nancy	F 1–3	8	8	11	11	100
		F 4	8	8	5	6	83
	Rabbit Slough	W			52	52	100
		F 1	8	8	4	4	100
		F 2	8	8	11	13	85
Reduced Left-Biased	Bear Paw	F 1–3	0	0	0	54	0
		F 4–5	1	1	0	21	0
		F 6–7	2	1	0	26	0
		F 8–18	2	2	0	141	0
		F 19–20	3	4	0	24	0
	Boot	F 1	2	2	0	4	0
		F 2	2	3	9	9	100
	Bruce	F 1–3	2	2	0	40	0
		F 4–6	6	6	0	25	0
		F 7	7	7	0	9	0
	Coyote	W			0	60	0
	Morvro	F 1–2	0	0	0	39	0
		F 3–6	1	1	0	67	0
		F 7	1	2	0	1	0
		F 8–12	2	2	0	70	0
		F 13	2	2	13	17	76
		F 14	2	2	3	12	25
		F 15	2	2	1	13	8
		F 16–17	3	3	0	18	0
		F 18	3	3	9	10	90
		F 19	4	4	15	30	50
		F 20	4	5	0	12	0
		F 21	6	6	16	36	44
		F 22	7	7	9	15	60
		F 23	8	8	4	14	29
	Wallace	F 1	2	4	1	11	9
	Whale	W			0	46	0
		F 1	2	2	0	3	0
	Zero	F 1	2	2	0	2	0
Reduced Right-Biased	L-shape	W			3	40	8
		F 1	2	2	0	14	0
		F 2	2	2	5	6	83
	Orphia	W			1	52	2
Reduced Unbiased	Dolomite	F 1	2	2	0	4	0
Spineless	Echo	W			9	44	21
		F 1	6	6	3	13	23
		F 2	6	6	6	15	40
		F 3	6	6	5	6	83
		F 4	6	6	6	7	86
	Kidney	F 1–3	6	6	0	38	0

See the text for definitions of parental phenotypes. The number stained and sample sizes were pooled among families from the same source and with the same parental phenotypes and frequencies of stained progeny.

Capture and preparation of wild-caught fry

Fry were captured using fine-mesh aquarium nets during June 2004 and fixed for 24 hours in chilled phosphate buffered saline (PBS) with 4% formaldehyde and transferred directly into 100% chilled methanol. The methanol was replaced with fresh chilled methanol after 1 h and 24 h. The 2004 fry samples were stored in a −80°C freezer in methanol for more than a year after fixation.

In vitro fertilization and rearing

We used methods described by Hagen (1967; see also Aguirre et al 2004) and crosses were performed in June 2005 and 2006. We stripped the eggs from ripe females and fertilized them with sperm from a male's chopped testes. Fry were reared in 10 cm diameter Petri dishes with daily water changes and antibiotic and fungicidal washes on the third day. The fry hatched after about eight days and began to eat brine shrimp nauplei (*Artemia salina*) two or three days later. They were reared in 60 L aquaria or 2.5 L plastic boxes until they reached Swarup's (1958) stage 31, but survival was poor in plastic boxes.

In situ hybridization

In situ hybridization of *Pitx1* was performed following standard methods (see Thisse et al 1993, Thisse & Thisse 1998, McMahon et al 2001). About a third of the fry from each family was preserved in PBS formalin at Swarup's (1958) stages 29, 30 and 31, when *Pitx1* is expressed in specimens with full pelvic structures (Shapiro et al 2004). Lab-crossed fry were chilled on ice in their Petri dishes to slow them down for staging. The best *Pitx1* staining results were obtained at stage 31, and we restaged fixed wild-caught fry and fry from crosses to select this stage for *in situ* hybridization.

The *Pitx1* DNA was transformed into competent *Escherichia coli*, isolated, and prepared for *in vitro* transcription and *in situ* hybridization using QIAprep® spin miniprep protocol. RNA was prepared with and without hydrolysis by sodium bicarbonate and sodium carbonate, which yielded similar results. The fry were rehydrated and bleached in a 6% solution of hydrogen peroxide (H_2O_2) in PBS with tween (PBT) under direct light with rocking for 3 h to reduce pigmentation. The fry were incubated for 30 minutes in a proteinase K ($10 \mu L/ml$) solution in PBT. An anti-digoxigenin (DIG) antibody specific for the DIG-labelled nucleotides was used for *in vitro* transcription to make the RNA probe. We treated the fry for 1 h in NBT/BCIP to visualize the antibody. Stained fry were observed and photographed under a dissecting microscope. Only specimens that stained for

Pitx1 expression in the jaws or branchial arches were scored for expression in the pelvic region (Shapiro et al 2004, 2006a). Pelvic limb bud staining appeared as a distinct spot, and the pelvic girdle sometimes appeared as a faint area of staining near the limb bud (Fig. 3). Hybridization with sense RNA, a control for non-specific hybridization, was distributed throughout the body.

Results

Almost all fry from the anadromous and five lake populations with full pelvic structures (Table 1, Fig. 4) expressed *Pitx1* in the pelvic limb bud during development (Table 2), but expression in the girdle was rarely visible (Fig. 3). Thus, we do not report frequencies of *Pitx1* expression in the pelvic girdle. Fry from populations with left-biased pelvic reduction produced variable frequencies of *Pitx1* staining in the pelvic limb bud. *Pitx1* was never expressed in the pelvic bud of wild-caught fry from Coyote and Whale populations, in which <2% of the wild-caught juveniles and adults have PS > 2 (Table 1). Similarly, progeny of parents with a PS of 0 or 1 from Bear Paw and Morvro, never expressed *Pitx1*. Progeny of parents with a PS of 2 from six populations with extreme left-biased pelvic reduction rarely expressed *Pitx1* in the limb bud. However, all nine progeny in a PS 2 × PS 3 cross from Boot (F2) expressed *Pitx1*, and some progeny in three out of eight PS 2 × PS 2 crosses from Morvro (F13–15) expressed *Pitx1* in the limb bud. Progeny from crosses using Morvro parents with higher pelvic scores usually expressed *Pitx1* (Morvro F18–23), but the progeny of parents with similar phenotypes from Bruce (F4–7) never did.

Pitx1 expression varied greatly among three populations with extreme pelvic reduction and right-biased or fluctuating asymmetry. Only one of 53 field-caught Orphia fry expressed *Pitx1* in the limb bud, which is expected from the frequency of pelvic phenotypes in this population (Table 1). Two crosses were performed using PS 2 parents from L-shaped Lake, and five of six progeny from L-shape cross F2 expressed *Pitx1* in the limb bud. Four of them expressed *Pitx1* only on the right, and the fifth expressed it more strongly on that side. About 8% of the field-caught fry from L-Shaped Lake expressed *Pitx1* in the limb bud, which is less than expected from the frequency of spine expression in the population (Table 1). None of only four fry from our Dolomite PS 2 × PS 2 cross expressed *Pitx1* in the limb bud. Overall, these results do not differ strikingly from those from left-biased populations.

None of 38 progeny from three Kidney (spineless) crosses expressed *Pitx1*. In contrast, *Pitx1* was expressed in about 20% of wild-caught Echo fry (Table 2), which is similar to the frequency of pelvic spine expression in the population (Table 1, PS = 7 or 8), and it varied from about 20 to 80% in the four Echo crosses.

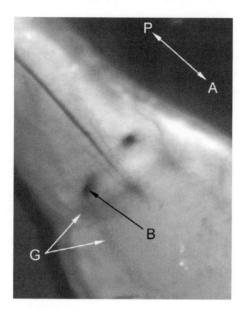

FIG. 3.   Ventral view of the pelvic region of a specimen with typically intense *Pitx1* staining in the pelvic limb bud (B) and faint staining of the pelvic girdle (G). A and P indicate anterior and posterior directions.

There appears to be a difference in *Pitx1* expression between the two spineless populations.

Discussion

Pitx1 specifies normal vertebrate hind limb development (Lancôt et al 1999, Szeto et al 1999, Minguillon et al 2005, Chang et al 2006), and *Pitx1* knockouts in the mouse have smaller hind limbs that tend to be larger on the left (Lancôt et al 1999, Marcil et al 2003). Previous studies of at least six threespine stickleback populations with pelvic reduction invariably implicated *Pitx1* in left-biased pelvic reduction (Cole et al 2003, Cresko et al 2004, Shapiro et al 2004). Nelson (1971) reported left-biased pelvic reduction in ninespine stickleback (*Pungitius pungitius*), and Shapiro et al (2006b) added genetic evidence that *Pitx1* causes pelvic reduction in this species. Shapiro et al (2006b) also demonstrated left-biased pelvic reduction in manatees (*Trichechus manatus*), implicating *Pitx1* in pelvic reduction of a distantly related mammal. Loss of *Pitx1* expression in the pelvis may be a widespread cause of pelvic reduction in vertebrates, and left-biased directional asymmetry of pelvic

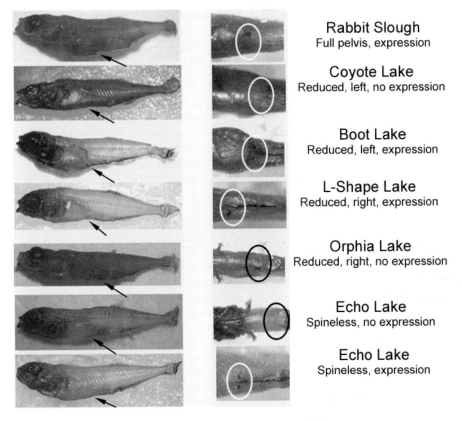

Rabbit Slough
Full pelvis, expression

Coyote Lake
Reduced, left, no expression

Boot Lake
Reduced, left, expression

L-Shape Lake
Reduced, right, expression

Orphia Lake
Reduced, right, no expression

Echo Lake
Spineless, no expression

Echo Lake
Spineless, expression

FIG. 4.   Lateral view (left) and ventral view of the pelvic region of selected specimens to show differences in *Pitx1* staining in specimens with and without pelvic reduction. Staining in the pelvic region is non-specific or represents *Pitx1* expression in the pelvic girdle.

vestiges indicates its involvement. Twenty of 27 Cook Inlet threespine stickleback populations with extensive pelvic reduction have left-biased directional asymmetry (Bell et al 2007). However, three had significantly right-biased pelvic reduction and another had fluctuating asymmetry, suggesting different genetic mechanisms for pelvic reduction in those populations.

   Wild-caught and lab-crossed stage 31 fry from six populations with full pelvic expression almost always expressed *Pitx1* in the pelvic limb bud. Inconsistent *Pitx1* staining in the pelvic girdle of these fry indicates that visualization of *Pitx1* expression in the pelvis is affected by developmental staging or some other methodological problem and was an unreliable indicator of *Pitx1* genotypes. Wild-caught fry from the two populations with the lowest mean pelvic scores never expressed *Pitx1* in the limb bud. In two populations with left-biased pelvic reduction, progeny of parents that lack the pelvis on one or both sides never express *Pitx1* in the limb bud. In six populations with extreme left-biased pelvic reduction, crosses using parents with small bilateral pelvic vestiges (PS = 2) also failed to express *Pitx1*. It appears that most stickleback with extreme pelvic reduction lack *Pitx1* expression. However, crosses from these populations sometimes produced some unexpected results. Some families from parents with relatively low pelvic scores expressed *Pitx1* in the limb bud, and others from parents with relatively high pelvic scores but no pelvic spines failed to express *Pitx1*. These exceptional results suggest that other genes beside *Pitx1* may cause extensive pelvic reduction, and that adults with extreme pelvic reduction may carry normal *Pitx1* alleles.

Previous studies of the genetics of pelvic reduction in threespine stickleback contrasted populations with either full pelvic expression or very extreme reduction (Cole et al 2003, Cresko et al 2004, Shapiro et al 2004, 2006). It is not surprising that our results are more complicated than theirs. Populations with extremely weak pelvic expression (i.e. Coyote, Whale) are probably fixed for *Pitx1* alleles that are never expressed in the pelvis, but *Pitx1* is probably polymorphic in more variable populations. In these populations, specimens with lower pelvic scores apparently are usually homozygous for *Pitx1* alleles that are silent in the pelvis. However, parents with slightly larger pelvic vestiges may carry a functional *Pitx1* allele, and perhaps their progeny express it because of recombination of alleles at multiple loci. It is surprising that even PS 2 parents (i.e. Morvro 13–15, L-shape F2) may carry *Pitx1* alleles that are expressed in the pelvis of their progeny. It is also interesting that all progeny from some parents with high pelvic scores (i.e. Bruce F4–7, Kidney F1–3) fail to express *Pitx1*. Shapiro et al (2004) detected four significant quantitative trait loci (QTL) besides *Pitx1* for pelvic reduction, and Peichel et al (2001) and Cresko et al (2004) reported a fifth one. Our results suggest relatively complicated interactions between *Pitx1* and other loci that contribute to pelvic reduction in other populations of *G. aculeatus*.

Right-biased and unbiased populations with extreme pelvic reduction were particularly interesting because their symmetry may indicate an unusual genetic basis for pelvic reduction. Nevertheless, wild-caught and lab-reared fry from these populations frequently failed to express *Pitx1*. These results may indicate that *Pitx2* is also silenced in these populations during pelvic development, reducing the frequency of specimens with a larger left pelvic vestige. Alternatively, other genes may cause the change in symmetry and also silence *Pitx1* expression in the developing pelvis. Identification of the loci with small effects on pelvic reduction and analysis of their expression patterns may resolve these possibilities.

The Kidney and Echo populations belong to a geographically restricted set of populations with high frequencies of pelvic spine loss (PS = 6) but full pelvic girdles (Bell & Ortí 1994). This truncated pelvic phenotype frequency distribution also suggests a different genetic architecture than that of other populations with pelvic reduction. However, *Pitx1* is expressed at variable frequencies in Echo crosses but never in Kidney crosses. These differences also may exist in left-biased populations with extreme pelvic reduction. As noted above, progeny from Bruce crosses using parents with similar pelvic phenotypes (Bruce F4–7) never expressed *Pitx1*, but crosses using parents with wide range of pelvic scores from Morvoro (F13–23) produced numerous progeny that expressed it. Pelvic spine development may be influenced by *Pitx1* or QTL on linkage groups 8 (Peichel et al 2001, Cresko et al 2004) and 4, which affect only pelvic spine development. Variation at these loci alone may account for pelvic spine loss without substantial reduction of the girdle in the Echo and Morvoro populations, but *Pitx1* appears to be involved in spine loss in the Kidney and Bruce populations, even when the girdle is not reduced.

Several genes, including *Tbx4*, *Pitx2*, *Fgf8*, *Fgf10*, *Shh*, *Hand2*, *HoxC10* and *HoxC11*, affect development t of the pelvic skeleton. For example, Thewissen et al (2006) attributed pelvic reduction in the spotted dolphin *Sternella attenuate* to failure of *Hand2* expression. Shapiro et al (2004) detected four factors besides *Pitx1* that influence pelvic reduction in one stickleback population, and Cresko et al (2004) interpreted partial complementation of pelvic phenotypes in interpopulation crosses to indicate the action of the same major locus but different minor loci for pelvic reduction in different populations. Thus, any of several genes might contribute to inconsistencies between *Pitx1* expression and the magnitude of pelvic reduction among populations and even families with the same parental phenotypes within populations. Our results add six populations with full pelvic structures that express *Pitx1* during pelvic development to two reported previously and nine populations with pelvic reduction that frequently fail to express *Pitx1* to two previous cases (Cole et al 2003, Shapiro et al 2004). *Pitx1* certainly is a common factor for pelvic reduction in *G. aculeatus* and other species, but there is more than one way to lose a vertebrate pelvis. Thus, pelvic reduction in threespine stickleback meets Jacob's (1977) expectation that selection may pull any of several spare genes off the shelf with which to tinker.

Acknowledgments

Several undergraduate and graduate students helped catch and rear stickleback. Laboratory facilities were provided by F. A. von Hippel and others in the Department of Biological Sciences, University of Alaska Anchorage, and by T. Lohuis, S. Jenkins and J. Crouse at the Alaska

238   BELL ET AL

Department of Fish and Game's Kenai Moose Research Center on the Kenai National Wildlife
Refuge. The *Pitx1* construct was provided by M. D. Shapiro and D. M Kingsley with support
of the Howard Hughes Medical Institute. We thank M. P. Travis and T. Sanger for help with
the figures. E. Londin, L. Mentzer and S. Jayaratna provided technical advice on *in situ* hybridization. KEE was supported by an REU supplement to NSF Grant DEB0322818, URECA
(Stony Brook University), and Howard Hughes Medical Institute Grant 52003746 to D. Bynum.
She participated in the 2005 Stickleback Molecular Genetics Workshop, Stanford University
Medical School. This research was supported by NSF DEB0322818 to MAB and F. J. Rohlf
and NIH grants to H. I. Sirotkin.

References

Aguirre WE, Doherty PK, Bell MA 2004 Genetics of lateral plate and gill raker phenotypes in a rapidly evolving population of threespine stickleback. Behaviour 141:
1465–1483
Bell MA 1987 Interacting evolutionary constraints in pelvic reduction of threespine sticklebacks, *Gasterosteus aculeatus* (Pisces, Gasterosteidae). Biol J Linn Soc Lond 31:347–382
Bell MA 1988 Stickleback fishes: bridging the gap between population biology and paleobiol-
ogy. Trends Ecol Evol 3:320–325
Bell MA, Ortí G 1994 Pelvic reduction in threespine stickleback from Cook Inlet lakes:
geographical distribution and intrapopulation variation. Copeia 2:314–325
Bell MA, Foster SA 1994 Introduction to the evolutionary biology of the threespine stickleback.
In: Bell MA, Foster SA (eds) The evolutionary biology of the threespine stickleback. Oxford
University Press, Oxford p 1–27
Bell MA, Ortí G, Walker JA, Koenings JP 1993 Evolution of pelvic reduction in threespine
stickleback fish: a test of competing hypotheses. Evolution 47:906–914
Bell MA, Travis MP, Blouw DM 2006 Inferring natural selection in the fossil record. Paleobiol-
ogy 32:562–576
Bell MA, Khalef V, Travis MP 2007 Directional asymmetry of pelvic vestiges in threespine
stickleback. J Exp Zool (Mol Dev Evol) 306B:189–199
Carroll SB, Grenier JK, Weatherbee SD 2005 From DNA to diversity: Molecular genetics and
the evolution of animal design, 2nd edn. Blackwell Publishing, Malden
Chang WY, Khosrow, Shahian F et al 2006 Conservation of *Pitx1* expression during amphibian
limb morphogenesis. Biochem Cell Biol 84:257–262
Cole NJ, Tanaka M, Prescott A, Tickle C 2003 Expression of limb initiation genes and clues
to the morphological diversification of threespine stickleback. Curr Biol 13:R951–952
Cresko WA, Amores A, Wilson C et al 2004 Parallel genetic basis for repeated evolution
of armor loss in Alaskan threespine stickleback populations. Proc Natl Acad Sci USA
101:650–655
Francis RC, Havens AC, Bell MA 1985 Unusual lateral plate variation of threespine sticklebacks
(*Gasteroseteus aculeatus*) from Knik Lake, Alaska. Copeia 1985:619–624
Gibson G 2005 The synthesis and evolution of a supermodel. Science 307:1890–1891
Gould SJ 1977 Ontogeny and phylogeny. Belknap Press of Harvard University Press,
Cambridge
Hagen DW 1967 Isolating mechanisms in threespine sticklebacks (*Gasterosteus*). J Fish Res
Board Can 24:1637–1692
Kingsley D, Zhu B, Osoegawa K et al 2004 New genomic tools for molecular studies of
evolutionary change in threespine sticklebacks. Behaviour 141:1331–1344
Lanctôt C, Moreau A, Chamberland M et al 1999 Hindlimb patterning and mandible develop-
ment require the *Ptx1* gene. Development 126:1805–1810

Logan M, Tabin CJ 1999 Role of *Pitx1* upstream of *Tbx4* in specification of hindlimb identity. Science 283:1736–1739

Marcil A, Dumontier E, Chanberland M et al 2003 *Pitx1* and *Pitx2* are required for development of hindlimb buds. Development 130:45–55

McMahon A, Bronner-Frasier M. Kingsley D 2005 Whole-mount in situ hybridization of vertebrate embryos. 27 January *http://kingsley.stanford.edu/Lab%20ProtocolsWEB%202003/Fish%20Methods/stickle_in_situs.htm*

Meyers EN, Martin GR 1999 Differences in left-right axis pathways in mouse and chick: functions of *Fgf8* and *Shh*. Science 285:403–406

Minguillon C, Buono JD, Logan MP 2005 *Tbx5* and *Tbx4* are not sufficient to determine limb-specific morphologies but have common roles in initiating limb outgrowth. Developmental Cell 8:75–84

Nelson JS 1971 Absence of the pelvic complex in ninespine sticklebacks, *Pungitius pungitius*, collected in Ireland and Wood Buffalo National Park region, Canada, with notes on meristic variation. Copeia 1971:707–717

Peichel CL, Nereng KS, Ohgi KA et al 2001 The genetic architecture of divergence between threespine stickleback species. Nature 414:901–905

Raff RA, Kaufman TC 1983 Embryos, genes, and evolution. Macmillan, New York

Reger RD, Pinney DS 1996 Late Wisconsin glaciation of the Cook Inlet Region with emphasis on the Kenai lowland and implications for early peopling. In: David NY, Davis WE (eds), Adventures through time: readings in the anthropology of Cook Inlet, Alaska. Cook Inlet Historical Society, Anchorage p 15–35

Shapiro MD, Marks ME, Peichel CL et al 2004 Genetic and developmental basis of evolutionary pelvic reduction in threespine sticklebacks. Nature 428:679–782

Shapiro MD, Marks ME, Peichel CL et al 2006a Genetic and developmental basis of evolutionary pelvic reduction in threespine sticklebacks. Erratum Nature 439:1014

Shapiro MD, Bell MA, Kingsley DM 2006b Parallel genetic origins of pelvic reduction in vertebrates. Proc Natl Acad Sci USA 103:13753–13758

Stern DL 1998 A role of *Ultrabithorax* in morphological differences between *Drosophila* species. Nature 396:463–466

Stern DL 2000 Perspective: evolutionary developmental biology and the problem of variation. Evolution 54:1079–1091

Swarup H 1958 Stages in the development of the stickleback *Gasterosteus aculeatus* (L.). J Embryol Exp Morph 6:373–383

Szeto DP, Rodriguez-Esteban C, Ryan AK et al 1999 Role of the *Bicoid*-related homeodomain factor *Pitx1* in specifying hindlimb morphogenesis and pituitary development. Gene Dev 13:484–494

Thewissen JGM, Cohn MJ, Stevens et al 2006 Developmental basis of hind-limb loss in dolphins and origin of the cetacean bodyplan. Proc Natl Acad Sci USA 103:8414–8418

Thisse C, Thisse B 1998 High resolution whole-mount *in situ* hybridization. The Zebrafish Science Monitor 5:8–9

Thisse C, Thisse B, Schilling TF, Postlethwait JH 1993 Structure of the zebrafish *snail 1* gene and its expression in wild-type, *spadetail* and *no tail* mutant embryos. Development 119:1203–1215

DISCUSSION

Duboule: Have you looked at the *situs* of the internal organs in these fishes? Do you get a complete *situs inversus*?

Bell: No. A number of people have suggested this, and I intend to do it.

Duboule: Is it known whether fishes have mutations in their global *situs*?

Bell: I know it occurs as a species-specific phenomenon. People have looked at it in zebrafish.

Hall: In most populations when you get the distribution of the whole range from 0 to 4 at a population level, have you looked at the progeny of individuals? Do you see the same distribution?

Bell: We have mostly focused on the populations with extreme pelvic girdle reduction. We usually do crosses using parents with a score of two (see Fig. 1 in my paper). The pelvis is represented by a little button of bone on each side. This parental phenotype mostly produces offspring with a score of two. Some progeny exhibit higher phenotypic scores, which would be expected from recombination. In addition, for convenience, we raised the families in 10% seawater because survival is better. There may be a bit of phenotypic plasticity based on water chemistry. This may be bumping the phenotypic score of progeny up a bit. If fish with a full pelvis are crossed you never get fish with pelvic reduction. This summer we did some crosses with parents with pelvic phenotypes of 2 × 2, 4 × 4, 6 × 6 and 8 × 8. They looked good, but the data are not available yet.

Hall: So it is not phenotypic plasticity at the individual level. It is plasticity at the population level.

Bell: No, phenotypic differences among populations reflect genetic divergence. They are evolutionarily labile.

Lieberman: Can you comment on the pathway/network of *Pitx1*, and what sort of hypotheses there are about how tinkering might be working on this.

Bell: I don't really do this work. We have looked at some of the papers concerning this developmental pathway, and it is not as frequently studied as the pectoral limb. *Tbx4* is in the pathway, as is an FGF. *Pitx2* is there but seems to function somewhat independently of the *Pitx1* part of the network. Clearly, there are more actors than that. Shapiro et al (2004) found four minor loci that affect pelvic expression. We mapped yet another minor locus that affects it (Cresko et al 2004), which Peichel et al (2001) had previously detected in another cross. *Pitx1* is the common part for tinkering, but there is also evidence from complementation crosses that different minor loci are important in different populations (Cresko et al 2004). When we do the complementation crosses we don't usually get the pelvis back, but we get slight enhancement, which suggests that minor loci are complementing in crosses between populations.

Lieberman: Elucidating these pathways can tell you a lot about what these mechanisms are.

Thesleff: *Pitx1* and *Pitx2* are expressed in teeth. The *Pitx2* mutants have no teeth. There is also the human Rieger syndrome with *Pitx2* mutations associated with missing teeth.

Bell: The mouse knockouts are not tissue specific. In the stickleback crosses in which there has been pelvic reduction, *Pitx1* doesn't light up in the pelvis but is still expressed in the pituitary, and most prominently in the branchial arches and in the jaws. If you get a good picture, they have brilliant purple lips. This is a good internal control, because if the fish expresses *Pitx1* in the jaws and branchial arches but not the pelvis, you can be sure the reaction worked and *Pitx1* is not expressed in the pelvis.

Wilkins: You said that dorsal spines are not under the control of *Pitx1* but they always seem to be lost with the loss of the pelvic architecture.

Bell: Not in our populations. They are almost always threespined. It is more likely that they will be fourspined than twospined, onespined, or no-spined.

Wilkins: Don't they decrease in size?

Bell: They tend to be small in populations that aren't sympatric with predatory fishes in general, even where there is no substantial pelvic reduction. When predatory fishes are absent it makes a big difference to the armour. When fabricating the pelvis becomes really expensive, this is when you see major elements disappear. The spines, pelvis and plates are a single functional unit. Clearly, the major factor in the plate reduction is a locus called Ectodyslpasin (Colosimo et al 2005), which is on a different chromosome than *Pitx1*. David Kingsley has a student working on spine number variation because their lake population does vary extensively for spine number. In the fossil record the spines are lost in sequence. There are other clues: dorsal spines never get below a certain size in the fossils and then they are present or absent. Lots of traits are being studied by a growing stickleback community, and it is clear that spine and pelvic girdle expression are genetically independent.

Stern: Recently Shapiro et al (2006) published a correction to their original study (Shapiro et al 2004). I think in the original study they accidentally used a sense probe. It was interesting, because the sense probe detected a pattern that was similar to the antisense probe. Is there any evidence that this other transcript is involved in regulating *Pitx1*? Have you looked for it?

Bell: It's not something we would do. We obtained our probe from Mike Shapiro and David Kingsley. They sent us the sense probe, and then I got an e-mail telling us not to use it. We used the antisense probe, with the sense probe as a control.

Hanken: You document the ontogeny of the spine in ancestral species. When one sees the reduced condition in extant adults, can you place these conditions along the ancestral ontogeny?

Bell: No, in the fossils (not the modern ones) you can place them along the ancestral ontogeny for the girdle. During development the spine appears very early, yet it is the first thing lost in fossils and modern populations. On the other hand, it develops in a different embryonic tissue, so the rules may be different for the spine and pelvic girdle.

Morriss-Kay: You referred to the interest ethologists had for a long time in sticklebacks. In fact, Nico Tinbergen worked on them as a boy. Have any of the ethologists described behavioural correlates to the changes in pelvic structure and spine structure?

Bell: This kind of work is being done in Felicity Huntingford's laboratory at the University of Glasgow. I don't know what the results are. It could go either way. If the armour works well, selection for predator avoidance behaviour might be relaxed, or if selection favours avoidance of predators, both armour and behavioural traits might diverge. The underlying genetic basis of a trait called boldness in stickleback is a multivariate measure of the willingness to engage with threatening objects. Lots of people are working on this. We are dabbling in it ourselves, but focusing on the genetics of brain morphology. We have found some major differences between populations.

Hallgrimsson: What is the covariation structure in spine length?

Bell: We have data from the fossils, and we have the data from at least one population that has been evolving rapidly. One of the other things we do in my lab is to study contemporary evolution. We studied a lake that was colonized by marine stickleback some time in the mid-1980s. When we first sampled in 1990, it looked more like a marine stickleback than any other lake stickleback we have sampled, because of body shape, armour and so on. By 2003 its phenotype was right in there among the lake fish for body shape, and for armour it was pretty close to the lake fish. We have these kinds of data so we could analyse them for trait covariation.

Budd: When I heard first about this work, I was interested in the fact that ionic concentration was correlated with changes. This idea of an environmental gradient plus selection suggests genetic assimilation to me. Do you think this is part of what is going on? The asymmetry effects would be interesting in this context: one wouldn't expect phenotypic plasticity to be expressed asymmetrically. Clearly, this would be a state like canalization by *Pitx1* genes if this were the case.

Bell: I have thought about phenotypic plasticity mostly as a nuisance in this context, because I'd really like them to breed true if it is heritable, so we can compare parents and progeny. I'm pretty sure it is there. I had a grant proposal on this turned down. One of the nice things about this organism is that you can do mass crosses: you can take the eggs from 10 or 20 females, the sperm from 10 or 20 males, and mix them in. Then you can do experiments on these. Now that we have a collaborator in Alaska, we can bring fish back to his lab and grow them in large tanks in lake water or artificial lake water with different salinity levels. These experiments can be done.

Budd: What is the generation time?

Bell: In the field it is two years. If you are not in a hurry, you can make it two years in the lab. You can push them to one year, or even six months. However, generation time is not as short as flies!

Jernvall: You have incredibly good time-resolution in your fossil sample, and there seems to be directional selection. Coming from work in mammals where we typically have a 1 million year time resolution, have you played with the data to see how your patterns would look if you increased time averaging by pooling individuals over longer time scales?

Bell: Yes. Clearly, greater time averaging should cause the variance of individual traits to go up (Bell et al 1987). Covariance between traits also goes up (Bell et al 1989). In one part of our stratigraphic section, there are spiny fish with a full pelvic girdle, and in another part of the section there are low-spined individuals without a pelvis. If you examine a large sample with the full range of pelvic and spine number variation from a single year, it contains all pair-wise combinations of phenotypes in the frequencies expected for random association (Bell et al 1989). However, when we started out, we found numerous correlations between traits using samples separated by thousands of years (Bell et al 1985). Even though there are no trait correlations among specimens within a year, correlations among samples result from pooling individuals across long time intervals within which multiple traits have evolved. Another issue is rates of evolution. Apparent rates go down over longer time scales in the fossil record. One consequence is that increasing the time interval over which rates of natural selection are measured biases the rates downwards, reducing the probability of eliminating drift as a cause of change. We couldn't falsify drift as a cause of change using 250 year time intervals (Bell et al 2006).

References

Bell MA, Baumgartner JV, Olson EC 1985 Patterns of temporal change in single morphological characters of a Miocene stickleback fish. Paleobiology 11:258–271

Bell MA, Sadagursky MS, Baumgartner JV 1987 Utility of lacustrine deposits for the study of variation within fossil samples. Palaios 2:455–466

Bell MA, Wells CE, Marshall JA 1989 Mass-mortality layers of fossil stickleback fish: catastrophic kills of polymorphic schools. Evolution 43:607–619

Bell MA, Travis MP, Blouw DM 2006 Inferring natural selection in a fossil threespine stickleback. Paleobiology 32:562–576

Colosimo PF, Hosemann KE, Balabhadra S et al 2005 Widespread parallel evolution in sticklebacks by repeated fixation of Ectodyslpasin alleles. Science 307:1928–1933

Cresko WA, Amores A, Wilson C et al 2004 The genetic basis of recurrent evolution: armor loss in Alaskan populations of threespine stickleback, Gasterosteus aculeatus. Proc Natl Acad Sci USA 101:6050–6055

Peichel CL, Nering KS, Ohgi KA et al 2001 The genetic architecture of divergence between threespine stickleback species. Nature 414:901–905

Shapiro MD, Marks ME, Peichel CL et al 2004 Genetic and developmental basis of evolutionary pelvic reduction in threespine sticklebacks. Nature 428:717–723

Shapiro MD, Marks ME, Peichel CL et al 2006 Genetic and developmental basis of evolutionary pelvic reduction in threespine sticklebacks. Erratum. Nature 439:1014

Using patterns of fin and limb phylogeny to test developmental– evolutionary scenarios

Michael I. Coates, Marcello Ruta* and Peter J. Wagner†

*Department of Organismal Biology & Anatomy, University of Chicago, Chicago, IL 60637, USA, *Department of Earth Sciences, University of Bristol, Bristol BS8 1RJ, UK and †Department of Geology, Field Museum of Natural History, Chicago, IL 60605, USA*

Abstract. Increasing fossil evidence surrounding the evolutionary origin of vertebrate limbs can be used to reconstruct the assembly of a limb ground-plan common to all tetrapods. The sequence of changes at the fin-to-limb transition can be compared to patterns of fin and limb ontogeny, and further comparisons can be made between phylogenetic changes at pectoral and pelvic levels. Such comparisons inform questions about the evolution of developmental autonomy (modularity). Limb evolution mostly concerns terminal additions and losses; from a developmental standpoint, these probably result either from minor adjustments to limb bud proportions or from the relative timing of gene expression or tissue growth. Evolutionary radiations of large clades are widely assumed to be marked by periods of rapid morphological diversification, raising further questions about the impact of restrictions imposed not only by ecology, but also by development and genetics. The early tetrapod data set is now large enough to allow initial tests of evolutionary inference to be conducted. New results are revealing novel patterns of evolutionary rate-change, encompassing the traditional notion of the fish-to-tetrapod transition and the root of the modern (crown-group) tetrapod radiation.

2007 Tinkering: the microevolution of development. Wiley, Chichester (Novartis Foundation Symposium 284) p 245–261

It might be assumed that 'Tinkering: the microevolution of development' concerns a subject beyond the reach of the fossil record. The incompleteness of the fossils themselves, let alone the fact that fossil-yielding rocks provide a patchy historical coverage of life on Earth, would seem to disqualify palaeontology from saying anything much about microevolutionary events. Fossils are thought to be much better suited for investigating macroevolutionary phenomena. Stanley (1998) characterized the problem succinctly, when he wrote (in 'Macroevolution') that '... paleontology has traditionally been valued primarily for its general documentation of large-scale rates, trends, and patterns of change. Its contribution has been

245

quite distinct from that of biology, which has focused on small-scale evolution within populations.' Furthermore, with a nod to Simpson (1944) he adopted Goldschmidt's (1940) distinction of a natural discontinuity between evolutionary process above and below the species level, and argued that this boundary marked the division between micro- and macroevolution. The present article is not concerned with supporting or challenging this notion of differences in evolutionary process at different hierarchical levels in taxonomy (although we return briefly to the question of evolutionary scale at the end of this article). However, the utility of fossils needs to be made clear, and, at least with reference to questions about ontogeny and phylogeny, it can be summarized under the following five headings (from Patterson 1981). Fossils can:

1. Suggest the phylogenetic sequence of character acquisition (and loss).
2. Expose non-homologies.
3. Refute phylogenetic inferences drawn from ontogenetic sequences.
4. Be used to define the minimal age of taxa.
5. Supply historical biogeographic data.

These points (with the exception of the fifth) set out the structure for the present article. The evolutionary scale of morphological changes addressed here is largely arbitrary. 'Tinkering' (dictionary definitions include 'experimental mending', and/ or 'meddling with parts') suggests small- rather than large-scale adjustments, and it is preferred to 'microevolution' for practical reasons, in so far as the anatomical areas of interest discussed here are confined to the pectoral and pelvic appendages.

Fossils can suggest the phylogenetic sequence of character acquisition (and loss)

Sampling taxa temporally close to common ancestors can improve reconstruction of ancestral character states and thus improve ideas about sequences of character gain and loss (Cunningham 1999). Available fossil material allows the reconstruction of fairly detailed hypotheses of the evolutionary transition from paired fins to limbs (Coates et al 2002, Shubin et al 2006). Such transformation scenarios depend upon the content of the tetrapod stem-group, which includes those extinct taxa that fall outside the modern tetrapods radiation (amphibians and amniotes plus all fossil relatives), and that are more closely related to tetrapods than they are to any other living group of lobe-finned fishes (Sarcopterygii), such as lungfishes or coelacanths (Jeffery 2001, Coates et al 2002). The modern tetrapod radiation constitutes the crown-group. Character-states from morphology-based phylogenies that occur at the node subtending this crown-clade equate with the hypothesized ground-plan of living tetrapods, or parts thereof such as limbs. The limbs of stem-group tetrapods, as they branch from successively more basal nodes on the phylogenetic tree (i.e. away from the crown), generally exhibit fewer and fewer characteristics shared with crown-group tetrapod limbs. Note that this does not imply that the appendages of such stem-group tetrapods are necessarily more primitive. Hypotheses of phylogeny for stem-group taxa tend to be depicted in a pectinate fashion, with a series of twigs connecting fossils in an ascending order towards a crown-taxon. While this ought to be merely a graphical convention, such depictions are sometimes indistinguishable from ancestor-descendent series (Panchen 1991). In the example of tetrapod limb evolution, some of the most specialized paired appendages of stem-group tetrapods belong to rhizodontids, an extinct clade of sarcopterygian fishes that branch from a point close to the base of the tetrapod stem-group (Jeffery 2001, Coates et al 2002, Shubin et al 2006) (Fig. 1A). These fins are clearly not ancestors of tetrapod limbs, but they do allow us to draw inferences about hypothetical ancestral conditions.

Eusthenopteron (Jarvik 1980) is the standard textbook example of a sarcopterygian fish exhibiting a pre-limb, fin-like appendage (Fig. 1B). *Eusthenopteron* is known from several well-preserved specimens, and the paired fins show the classic pattern of one proximal to two distal bones (resembling the relation of humerus to radius and ulna, or femur to tibia and fibula). However, this pattern is widespread among the finned members of the tetrapod stem-group, and it may turn out to be not only conservative, but also primitive. In fact, there remains significant uncertainty

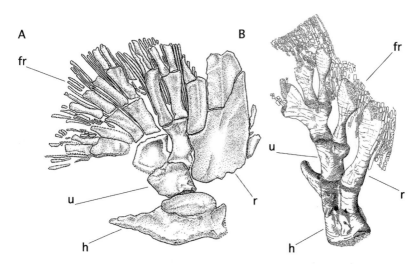

FIG. 1.   (A) pectoral fin of the rhizodontod stem-group tetrapod *Sauripteris* (after Daeschler & Shubin 1998). (B) Pectoral fin of the tristichopterid stem-group tetrapod *Eusthenopteron* (after Andrews & Westoll 1970). Leading edge of fins to right of figure, both shown in dorsal view; fins not drawn to scale. Abbreviations: fr, fin-rays; h, humerus; r, radius; u, ulna.

about the primitive condition for the tetrapod stem-group, and sarcopterygians as a whole (Ahlberg 1989, Coates 2002). Aside from the example of rhizodont fins, variation from the *Eusthenopteron* pattern is apparent only towards the apex of the stem-group, in fishes termed 'panderichthyids' (although these may be a grade rather than clade). In panderichthyids (Fig. 2), the endoskeleton of the shoulder girdle is enlarged, the dermal skeleton reduced, and the most proximal bone of the fin skeleton, the humerus, resembles the humeri of early tetrapod limbs (Coates et al 2002, Shubin et al 2006). However, the distal parts of panderichthyid appendages remain essentially fin-like, consisting of endoskeletal bones radiating out towards the periphery (radials), and dermal skeletal fin rays (Fig. 2A).

Fossils can expose non-homologies

Simulation studies (Huelsenbeck 1991) suggest that the primitive character state combinations of fossil taxa can reveal homoplasy where extant taxa alone indicate homology. This is akin to the long-branch problem (Felsenstein 1978) and although morphological data theoretically could be immune to this problem, constraints on possible forms coupled with limits of human perception prevent this ideal (Wagner 2000). The functional and architectural constraints limiting independence among

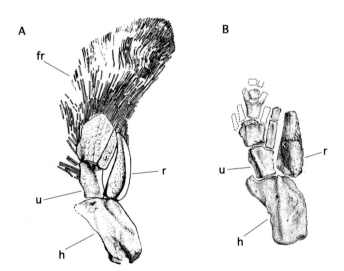

FIG. 2.   (A) Pectoral fin of the panderichthyid stem-group tetrapod, *Panderichthys* (after Vorobyeva 2000); ventral view. (B) Pectoral fin of elpistostegid stem-group tetrapod *Tiktaalik* (after Shubin et al 2006); dorsal view, fin-rays present but not shown. Leading edge of fins to right of figure; fins not drawn to scale. Abbreviations: fr, fin-rays; h, humerus; r, radius; u, ulna.

homologies on complex organisms such as vertebrates exacerbate this. Thus fossils should provide phylogenetic information that is lost among modern taxa. Particular details of the various fossil fin-like and limb-like appendages, and the order of changes implied by phylogenetic tree shapes, can be obtained elsewhere (Coates et al 2002, Shubin et al 2006). The aim here is to discuss the more general features of early limb evolution, and several of these features concern homoplasy. From a tetrapod perspective, such convergences or parallelisms appear as repeated experiments producing limb-like conditions. Returning to the pectoral fins of rhizodontid stem-tetrapods (Fig. 1A) (Jeffery 2001), convergent features include the enlargement of the endoskeletal girdle, and, away from the body wall, the fin skeleton expands anteriorly and posteriorly to support a wide paddle, resembling the 'hand-plate' (autopodium) of limbs (compare Figs 1A and 3A). The radials of the fin skeleton superficially resemble digits, although, unlike digits, they branch as they extend towards the fin perimeter. However, no known rhizodontid fins show evidence of fin-ray loss. All examples have well-developed rays, more or less like those of a cod or a zebrafish. The implication is that the pectoral fin buds of the embryos of these extinct fish presumably had distal or apical ectodermal folds enclosing the developing fin-rays, just as in modern bony fishes. In turn, this suggests that, unlike tetrapod limb buds, these ancestral fin buds included a mixed population of cells: some from the mesoderm for internal skeleton and muscle,

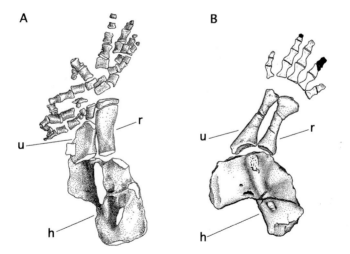

FIG. 3. (A) Pectoral limb of the stem-group tetrapod, *Acanthostega* (after Coates et al 2002). (B) Pectoral limb of stem-group tetrapod *Greererpeton* (adapted from Coates et al 2002, and refs therein). Both limbs shown in dorsal view. Leading edge of limbs to right of figure; limbs not drawn to scale. Abbreviations: h, humerus; r, radius; u, ulna.

250 COATES ET AL

and others from the neural crest. This is because, as far as is known, all components of the dermal skeleton (such as fin rays) are neural crest derivatives (Thorogood 1991, Hall 2000).

Fin-ray loss, however, is not the sole preserve of tetrapods. Looking beyond the tetrapod stem-group, lungfish show independent instances of fin-ray loss in muscular fins with an extended internal skeleton running to the outermost extremity. *Protopterus*, the African lungfish, retains only a short fringe of rays along the trailing edge of the pectoral fins, and none on the pelvic fins (Goodrich 1930, Coates 1994, Richardson et al 2004). Moreover, the pelvic fins are more robust than the pectoral fins, and are used for (mostly bipedal) locomotion along the substrate, with an alternating walking gait (personal observation M. I. Coates). Again, this resembles characteristics of tetrapods, and parallels the inferred sequence of functional changes in the fish-to-tetrapod transition: not only a shift from axial to appendicular musculo-skeletal drive, but also a change from pectoral level to pelvic level as the source of locomotor thrust.

Thus far, no fossil stem-tetrapod is known with fin-rays on the pectoral appendages and none on the pelvics. Similarly, all limbed early tetrapods have pectoral and pelvic appendages with series of digits arranged anteroposteriorly. This might be the defining feature of a tetrapod limb: as a morphological marker it is the outstanding indicator that, from a developmental perspective, the distal end of tetrapod limbs is a distinct region (Coates et al 2002, Richardson et al 2004). However, despite the absence of any fossil tetrapod with pectoral fins and pelvic limbs (or vice versa), it is apparent that the evolutionary transformation from fins to limbs concerns two events. Changes at pectoral and pelvic levels were, to some degree, mutually independent. Comparison of the earliest fossil tetrapod hindlimbs and forelimbs shows that the pelvic appendages are more conventionally limb-like than the pectoral appendages. There is a lag in the evolution of these serial homologues: pectoral limb evolution does not appear to occur in concert with pelvic limb evolution. By coding for pectoral and pelvic limbs as if they were separate, terminal taxa in a phylogenetic analysis, the resultant tree shows that hindlimbs of more primitive taxa cluster with forelimbs of more derived taxa (Coates et al 2002). Thus, forelimb evolution converges on patterns first seen in hindlimbs.

Fossils can refute phylogenetic inferences drawn from ontogenetic sequences

Although general relationships probably exist between ontogenetic sequences and phylogenetic sequences, ontogenetic sequences can change over phylogeny. This might be especially important in complex organisms such as vertebrates. Final points to add to this summary of early limb evolution include the now widely known discovery that the earliest tetrapod limbs exhibited patterns of six to eight digits (Fig. 3A). Furthermore, in at least one instance (*Ichthyostega*) there is evidence from several specimens that the seven-digit (four stout and three slender) pattern, at least in the hindlimb, was a species-specific condition (Coates 1991). Digits precede the evolution of wrists and ankles, and these complex joints continue to elaborate through phylogeny, adding more bone or cartilage components as the digit number is reduced to five or fewer (Coates et al 2002). As regards ways in which fossils inform questions about ontogeny and phylogeny, tetrapod limb development does not show evidence of either digit number reduction or fin ray loss. Fossil (rather than developmental) data set the agenda of evolutionary change: the pattern of historical morphologies and the earliest and/or most primitive manifestation of a morphological novelty. By applying a parsimony model to developmental evolution, it is possible to reconstruct some of the evolutionary steps involved in the assembly of the developmental network underpinning morphogenesis of structures in extant taxa.

There is some overlap here with questions addressing the origin of developmental modules: these are defined as compartments of organisms that display varying degrees of independence and can be the object of evolutionary transformation (adapted from Wagner 1996). The developmental module is a slippery concept, and the definition above is little different from working definitions of characters in morphology-based phylogenetic analyses. Modules and characters are subunits of individuals, and both show hierarchical organisation. Character evolution can be explored through phylogeny, and the same is true for module evolution. Tetrapod limb buds (Cohn & Tickle 1996) represent a classic set (group or class) of developmental modules, and the history of their study has been influenced by the notion that they represent a more-or-less discrete iteration of the tissues, morphogenetic processes and patterning events employed elsewhere in a developing vertebrate embryo. With regard to evidence of developmental evolutionary tinkering, or even *bricolage* (construction from bits and pieces), it is hardly surprising that limbs and limb buds incorporate many features of the median fins, which are known to occur earlier in phylogeny (Coates 1994). This is not an argument that paired fins and limbs represent transformations of median fins, in the sense that limbs can be conceived of as singular transformations of hypothetical ancestral paired fins. The relationship is, nevertheless, one of transformational iterative homology (interpretation of a structure as one of a set of structures), based on shared features of development (ontogeny) (Panchen 1992). The same kind of homology hypothesis applies to the relation of forelimbs to hindlimbs: hindlimbs have always been different to forelimbs, and one is not a transformation of the other (Coates & Cohn 1998). Similarly, from their evolutionary origin it is highly probable that paired fins and limbs lacked the bilateral symmetry of midline outgrowths (e.g. median fins). It is not necessary to invoke an archetypical paired limb, or an archetypical limb-development module. Instead, it is more productive to think of vertebrate fins and limbs as the result of different modules (and sub-modules, cf. regionalization of limb buds, Richardson et al 2004) evolving from different sets of initial conditions.

Phylogenies including fossils allow the reconstruction of hypotheses of ancestral conditions, and in so doing provide unique clues about developmental features that might have been acquired and discarded along the way. The idea that observed developmental diversity is contingent upon phylogeny is as old as any discussion of biological evolution. For vertebrate embryos, the classic illustration of this is the pharyngula: the 'phylotypic stage' (Duboule 1994) in which tetrapods display serial bulges of pharyngeal pouches, just as in fish embryos, and in both instances assumed to be retained from a common ancestral condition. Much the same might be said of the apical ectodermal ridge of limb buds: in this case the vestige of the ectodermal fold that houses the developing rays in fin buds. In fact, the apical fold and some of its signals (Niswander et al 1993, Sun et al 2002) is probably the most general and primitive part (or module) of all fin and limb developmental systems, given that the earliest fossil fins (median and paired) are preserved as keels and rays of the dermal skeleton without any indication of endoskeletal input (Coates 1994).

Fossils can be used to define the minimal age of taxa

Scenarios of developmental evolution are not confined to patterns of structural transformation; they also concern rates of change. The best known of these is the 'Cambrian explosion' (Gould 1989) for which various theories have been offered as an explanation of the sudden appearance of all living phyla within a geological instant. Resultant debate has (usefully) underscored key issues such as the importance of taxonomy and systematic biology (Briggs et al 1992), as well as the necessity of taking into account the quality of the fossil record itself (Wagner 1995). These topics, and especially the last, have tended to be neglected relative to searches for transitional forms and other kinds of fossil ancestor. With regard to the subjects of the present essay, the data set on early tetrapods is now large enough to allow wide-reaching analyses of morphological changes and diversification rates. Such topics move from supposedly micro- into macroevolutionary territory (Stanley 1998), but are directly relevant to discussions of developmental evolution.

Numerous major evolutionary radiations are marked by rapid diversification of new morphotypes (Wagner 2001), and the tetrapods are a good test case. Here, the origin of a new body-plan is associated with the invasion of new habitat (the terrestrial environment) and the crown-group divergence of amphibian and amniote clades. Crucially, two aspects of tetrapod evolution allow us to contrast predictions of instrinsic constraint (i.e. developmental or genetic) with ecological restriction (i.e. filling of general ecospace) models for reducing rates of morphological change (Valentine 1980). The first of these aspects is that the results of phylogenetic analyses indicate that post-Devonian tetrapods evolved from a single Late Devonian/Early Carboniferous taxon (Ruta & Coates 2007). This phylogenetic 'bottleneck' (*sensu* Jablonski 2002) yields a second radiation into similar (i.e. semi-aquatic) ecospace. To use Gould's (1989) metaphor, this gives us an opportunity to see a re-run of the evolutionary tape. Therefore, if ecology alone is responsible for rates of morphological change, then Early Carboniferous rates ought to mimic Devonian rates. Alternatively, if intrinsic constraints accumulated in the interim, then rates of morphological change might be lower. The second aspect in question is the (further) diversification of stem- and basal crown-amniotes in the Late Carboniferous. This represents a second invasion into yet more new ecospace: fully terrestrial habitats. If ecological restrictions affect rates of morphological change more strongly than intrinsic constraints, then the expected pattern would be a further episode of high rates of change. Conversely, if intrinsic constraints accumulated prior to this radiation, then there might be no increase in rates of change, although taxon disparity would increase steadily with subsequent diversification.

If the model of tetrapod phylogeny is correct (Ruta & Coates 2007), then the following result characterizes the sequential evolutionary radiations of early tetrapods: rates of change in the Permian were approximately *10 times* lower than in the Devonian (Ruta et al 2006). Tetrapods display a striking decrease in amounts of evolutionary change over time. It could be argued that this result is an artefact of poor sampling at the start of the tetrapod fossil record, but basal branches of the phylogenetic tree would have to be extended 10-fold (into the past) to dissipate differences between Devonian and Permian rates of evolutionary change. Furthermore, the most dramatic change is between the Devonian and the Early Carboniferous; there appears to be some stabilisation later (less significant difference in rates between the late Carboniferous and Permian). This initial burst of morphological evolution could easily represent 'relaxation' of both kinds of constraint: reduced ecological restrictions and reduced intrinsic, developmental and/or genetic constraints. However, the decreased rates in the Early Carboniferous are difficult to interpret other than as a result of increased intrinsic constraint. If there was a phylogenetic bottleneck, then the ecospace should have been nearly as available as it was in the Late Devonian.

Whether we can ever obtain similar results for partitions of the data, such as limb or fin morphologies, is doubtful, because the set of fossil material is likely to remain too fragmentary (at least for the foreseeable future). Even so, it is relevant to the present discussion to note that the origin of tetrapod limbs emerges from within this period of exceptionally rapid change. However, small-scale developmental changes might have large-scale anatomical effects. Perception of evolutionary scale is dependent upon the kind of data examined—confounding distinctions between micro and macroevolutionary change (Stanley 1998). The origin of digited limbs, relative to the nearest fins (phylogenetically speaking) might relate to not much more than a couple of subtle shifts in developmental patterning, whereas the origin of wrists and ankles could require considerable change at a developmental regulatory level, but appear as only one or two character-state changes in the morphological tree.

Acknowledgements

This work was funded by the faculty research fund, Pritzker School of Medicine, University of Chicago (MIC), a John Caldwell Meeker Research Fellowship, Field Museum Geology Dept., Chicago (MR), and NSF grant EAR 0207874 (PJW).

References

Ahlberg PE 1989 Paired fin skeletons and relationships of the fossil group Porolepiformes (Osteichthyes: Sarcopterygii). Zool J Linn Soc 96:119–166
Andrews SM, Westoll TS 1970 the postcranial skeleton of *Eusthenopteron foordi* Whiteaves. Trans R Soc Edinb 68:207–329
Briggs DEG, Fortey RA, Wills MA 1992 Morphological disparity in the Cambrian. Science 256:1670–1673
Coates MI 1991 New palaeontological contributions to limb ontogeny and phylogeny. In: Hinchliffe JR, Hurle JM, Summerbell D (eds) Developmental patterning of the vertebrate limb. NATO ASI Series A Life Sciences Vol 205. Plenum Press, New York p 325–337
Coates MI 1994 The origin of vertebrate limbs. In: Akam M, Holland P, Ingham P, Wray G (eds) The evolution of developmental mechanisms. Development 1994 supplement. The Company of Biologists Limited, Cambridge p 169–180
Coates MI, Cohn MJ 1998 Fins, limbs, and tails: outgrowth and axial patterning in vertebrate evolution. Bioessays 20:371–381
Coates MI, Ruta M, Jeffery JE 2002 Fins to limbs: what the fossils say. Evol Dev 4:390–401
Cohn MJ, Tickle C 1996 Limbs: a model for pattern formation within the vertebrate body plan. Trends Genet 12:253–257
Cunningham CW 1999 Some limitations of ancestral character-state reconstruction when testing evolutionary hypotheses. Syst Biol 48:665–674
Daeschler EB, Shubin N 1998 Fish with fingers? Nature 391:133
Duboule D 1994 Temporal colinearity and the phylotypic progression: a basis for the stability of a vertebrate Bauplan and the evolution of morphologies through heterochrony. In: Akam M, Holland P, Ingham P, Wray G (eds) The evolution of developmental mechanisms. Development 1994 Supplement. The Company of Biologists Limited, Cambridge p 135–142
Felsentstein J 1978 Cases in which parsimony or compatability methods will be positively misleading. Syst Zool 27:410–410
Goldschmidt, R 1940 The material basis of evolution. Yale University Press, New Haven
Goodrich ES 1930 Studies on the structure and development of vertebrates. Macmillan, London
Gould SJ 1989 Wonderful life. WW Norton, New York
Hall BK 2000 Evolution of the neural crest in vertebrates. In: Olsson L, Jacobson C-O (eds) Regulatory processes in development. Wenner-Gren International Series Vol 76. Portland Press, p 101–113

Huelsenbeck JP 1991 When are fossils better than extant taxa in phylogenetic analysis? Syst Zool 40:458–469

Jablonski D 2002 Survival without recovery after mass extinctions. Proc Natl Acad Sci USA 99:8139–8144

Jarvik E 1980 Basic structure and evolution of vertebrates. Academic Press, London

Jeffery JE 2001 Pectoral fins of rhizodontids and the evolution of pectoral appendages in the tetrapod stem-group. Biol J Linn Soc 74:217–236

Niswander L, Tickle C, Vogel A, Booth I, Martin GR 1993 FGF-4 replaces the apical ectodermal ridge and directs outgrowth and patterning of the limb. Cell 75:579–587

Panchen AL 1991 The early tetrapods: classification and the shapes of cladograms. In: Schultze H-P, Trueb L (eds) Origins of the higher groups of tetrapods: controversy and consensus. Cornell University Press, Ithaca p 110–144

Panchen AL 1992 Classification, evolution, and the nature of biology. Cambridge University Press, New York

Patterson C 1981 Significance of fossils in determining evolutionary relationships. Annu Rev Ecol Syst 12:195–223

Richardson MK, Jeffery JE, Tabin CJ 2004 Proximodistal patterning of the limb: insights from evolutionary morphology. Evol Dev 6:1–5

Ruta M, Coates MI 2007 Dates, nodes, and character conflict: addressing the amphibian origin problem. J Syst Palaeontol 5:69–122

Ruta M, Wagner PJ, Coates MI 2006 Evolutionary patterns in early tetrapods. I. Rapid initial diversification followed by decrease in rates of character change. Proc R Soc Lond B Biol Sci 273:2107–2111

Shubin NH, Deaschler EB, Jenkins FA Jr 2006 The pectoral fin of *Tiktaalik roseae* and the origin of the tetrapod limb. Nature 440:764–771

Simpson GG 1944 Tempo and mode in evolution. Colmbia University Press, New York

Stanley SM, 1998 Macroevolution, pattern and process. pbk. edn. The John Hopkins University Press, Maryland

Sun X, Mariani FV, Martin G 2002 Functions of FGF signalling from the apical ectodermal ridge in limb development. Nature 244:492–496

Thorogood P 1991 The development of the teleost fin and implications for our understanding of tetrapod limb evolution. In: Hinchliffe JR, Hurle JM, Summerbell D (eds) Developmental patterning of the vertebrate limb. NATO ASI Series A Life Sciences Vol 205. Plenum Press, New York p 347–354

Valentine JW 1980 Determinants of diversity in higher taxonomic categories. Paleobiology 6:444–450

Vorobyeva, EI 2000 Morphology of the humerus in the Rhipidistian Crossopterygii and the origin of tetrapods. Paleontol J 34:632–641

Wagner G 1996 Homologues, natural kinds and the evolution of modularity. Am Zool 36:36–43

Wagner PJ 1995 testing evolutionary constraint hypotheses with early Paleozoic gastropods. Paleobiology 21:248–272

Wagner PJ 2000 Exhaustion of cladistic character states among fossil taxa. Evolution 54:365–386

Wagner PJ 2001 Constraints on the evolution of form. In: Briggs DEG, Crowther PR (eds) Palaeobiology II. Blackwell, Oxford, p 154–159

DISCUSSION

Carroll: You have provided an effective look at the fish potential ancestry in structural and phylogenetic genetic terms. Unfortunately, when we step over this magic boundary between the Devonian and the Carboniferous, we walk into an abyss that has been called Romer's gap. Dr Romer spent much of his life looking for Lower Carboniferous tetrapods with very little success. I am working with the oldest tetrapod locality from the Carboniferous. It is extremely frustrating: we have lots of fossils, but almost all of them individual bones. We have four different kinds of humerus, two or three kinds of femur and a diversity of shoulder girdles. So we know that diversity was increasing just after the end of the Devonian, but we don't see much continuity between the Devonian and Lower Carboniferous tetrapods. There is only one putative sister taxon of the later forms, *Tulerpeton*, which may have connections with a specific lineage of anthracosaurs that may be closer to amniotes than modern amphibians. Basically, we have appalling ignorance of what happens next. This isn't necessarily a biological problem, but more likely a geographical and geological one. We simply haven't found the beds to go on with this story. Across the border there is gradual accumulation of about 19 lineages, roughly the same number as placental mammals that radiated after the extinction of dinosaurs. It is now impossible to establish definitive relationships between any of these lineage because of the paucity of the fossil record.

Coates: I disagree. I think we have a lot of data. We have a robust minimum date for the crown node (meaning the evolutionary divergence of amphibians from amniotes), but I concede that this is influenced by information from few fossil localities. One of these, East Kirkton in Scotland, is significant because it includes the earliest members of most of the early tetrapod clades. Consequently, whatever (evolutionary) tree shape we reconstruct, there are few tree topologies that will not place the tetrapod crown radiation as having occurred by the date of the East Kirkton fauna.

Carroll: In fact, it probably occurred by the Upper Devonian, or certainly very early in the Carboniferous.

Coates: A Late Devonian date for the tetrapod crown radiation is unlikely, because we should also take into account what is known about the emergence of terrestrial ecosystems. Changes affecting the atmosphere and terrestrial environment throughout Devonian are radical. The point is that terrestrial ecosystems that could support a vertebrate community occur for the first time at the very end of the Devonian. Suddenly (in terms of geological time), there was a new environment that provided the conditions that might have allowed a tetrapod evolutionary radiation, from which today's amphibian and amniote lineages arose. We can speculate that the crown node could have been deeper (in time), but the 'one (Devonian) putative sister taxon of the later forms, . . . which may be closer to amniotes than modern amphibians' concerns an out-dated hypothesis (Lebedev & Coates 1995). All of the data used for that analysis are now included within a much larger data set (Ruta et al 2003), and the earlier hypothesis has been superseded. We need to do more sophisticated treatments of the kinds of material we have concerning the fish–tetrapod transition, rather than saying we can't do anything without another fossil.

Carroll: After the whatcheerids, the oldest known fossil is an aïstopod, a group that already had no limbs and a fantastically modified skull. The next is a *Casineria* which is a putative amniote. We are only about 10–20 million years into the Carboniferous. Then there are the adelogyrinids, which are another totally limbless group, with the skull totally different from aïstopods. These are as far apart as we can get among primitive tetrapods, within 20 million years of the end of the Devonian. Many cladograms have been presented, but it is difficult to make reliable trees when you have such enormous structural disparity in the oldest known members of these groups.

Coates: All this disparity: agreed, there is a lot of change evident among tetrapods in the Lower Carboniferous, but cladograms from current data sets (used and developed by different research groups) converge on only two general branching patterns, both of which imply that the root of the modern tetrapod family tree lies within the Carboniferous. So, there is some consensus among these hypothesized trees. Furthermore, the first articulated remains of a tetrapod are of a conservative, almost salamander-like, form (Clack 2002a), at around 12 million years into the Carboniferous, and the limbless form co-occurs with other early tetrapods at about 25 million years into the Carboniferous (Clack 2002b).

Hall: We tend to think of homoplasy as noise and something we need to get rid of. But the notion of homoplasy as a null hypothesis for developmental change is an intriguing one. We could use this to design experiments with extant taxa. Could you elaborate this more?

Coates: I can give you an example of work completed recently with Marcello Ruta and Peter Wagner (Wagner et al 2006). The statistical method employed (by Peter) is one that has been used to estimate numbers of unseen species, and even the number of words that Shakespeare knew. Instead of re-sampling batches of taxa (or sections of text), this method can be applied to morphological data, with the aim of searching for 'character exhaustion', in much the same way that availability of a limited vocabulary results in cliché-laden, repetitive, speech patterns. If we sample the morphological character sets used by amphibian and amniote lineages, then it becomes apparent that amphibians are exhausted much more quickly than the amniotes. This is something that, perhaps, we get a feel for intuitively when looking at living amphibia. If taxa keep resampling the same kinds of morphological characteristics, then what we are talking about, effectively, is homoplasy. This takes us back to tinkering as a metaphor for reusing, developmentally, what is on-the-shelf.

Donoghue: I have a student, Graeme Lloyd, who has been mining cladistic data sets. He finds this to be a general pattern. There is early exhaustion in almost every data set.

Coates: The point here is that we can compare two clades directly. You have to be careful to examine taxa branching from a common node. I am interested in shark evolution, and can list many differences between a fossil shark and a living shark, but if we are to look at that side of the gnathosome radiation, a modern shark is a lot less different from a primitive shark than you, me or a duck compared to a very primitive osteichthyan fish (of which we are the equivalent subset). I don't know whether this character exhaustion occurs in stepwise pulses. Perhaps this is a pulse we are looking at with the tetrapods. This is why I was anxious to say that I am cautious about edge effects.

R Raff: Going back to tinkering, one of the intriguing things about the tetrapod limb is that you have two periods of transition going on. One is in that transformation from a fin into something that looks like a tetrapod limb. The other is the change from that polydactorous tetrapod limb to something that is pentadactalous and looks like a modern hand. It would be interesting to know something about the rate at which these things happen, and it would be really interesting to know something about the kind of reorganization that might be going on. Are those extra fingers all equivalent to each other? Are they simply lost? What has happened in that transformation?

Coates: When we reduce our data set down to look at early limbs, the sample size becomes too limited for analyses of rates. When Devonian tetrapod polydactylous limbs were first reported, Cliff Tabin proposed that there was an underlying pentate plan: you could have no more than five kinds of digits. I challenged him on this, and argued that the eight-digit fore limb of the Devonian tetrapod *Acanthostega* shows only a generalized set of 'fingers'. However, in each of the seven or eight examples known of the hind limb of *Ichthyostega* (contemporary with *Acanthostega*) there is a distinct pattern of four large posterior digits and a cluster of three small anterior digits: already there are different 'kinds' of digit.

Lieberman: Going back to what Jim Hanken mentioned earlier and Jukka Jernvall's analysis of the morphological apace of teeth, has anyone tried to figure out the total range of possibilities for the limb, and then map out which actually exist? What we have is a subset of some theoretical possibility. What is not there may be just as interesting as what is there.

Coates: Nigel Holder (1983) attempted this, and illustrated a series of forbidden morphologies. The natural historians leapt on it: for every rule, there seems to be a naturally occurring exception.

Hall: Dan Lieberman is presumably thinking of some defined morphospace. How many different phalanges can you have, for example?

Coates: That's a good question, and software for such morphometric analyses is now widely available.

Morriss-Kay: We have one mouse mutant in which we seem to get the maximum number of digits that a mouse limb bud is probably capable of making. It has six to nine digits in each limb. They are all triphalangeal. There are occasional little, narrow digits. Rudy Raff asked about the genetic changes that might have led from polydactyly to the pentadactyl limb. This mutant has a deletion that is close to Ihh. It has ectopic expression of Ihh all the way round just below the AER, not going into the region of Shh. We also crossed this mouse with the Shh knockout and showed that when there is Ihh all the way round, including into the Shh expression region, the result is that even more digits form. It could be that whatever this element is that is being deleted in these mice, it is something that is essential for restricting the number of digits to the pentadactyl pattern.

Duboule: Zakany et al (1997) showed that if you decrease the number of doses of Hox genes in a stepwise fashion, for example by removing an entire complex, you reduce the number of digits going from pentadactyly to adactyly via a phase of polydactyly. Simply by playing with the doses we can repeat this transition backwards. The number of digits that you can have is solely dependent on the space available. If you increase the space in the bud you don't increase the thickness of digits, but rather their number. The size of the condensation is genetically fixed in a given species. There are many different ways of increasing the space. One is what Gillian Morriss-Kay described: if you add signalling molecules, this may increase the space and the result is more digits. But keep in mind that regardless of the number of digits you obtain, they will likely be different from each other. The method used to increase the space will necessarily induce an asymmetry. In the case that it does not, the animal will be a dead animal. You can produce a collection of symmetrical digits, but on a genetic background the animal cannot stand.

Hall: The maximum number of digits in chick embryo is 10 or 12 in the *Talpid* mutant. The maximum number (I know of) in cats is 23 or 24 digits per limb, resulting from duplication of the digital arcade along the proximo–distal limb axis, giving 11–13 digits per row.

Coates: Sewall Wright had a strain of guinea-pigs with 12 per limb.

Budd: One thing you touched on is the idea that we can use fossils to trace the development of functional complexes in morphology, which I regard as being important organizers for development. The example you showed was the idea of the hind/forelimb difference. Clearly the hind limbs predate what happens in the forelimbs. This tells us something important when we try to compare the developmental systems associated with these two rather similar structures.

Coates: I agree.

Budd: As anyone who has worked on stem and crown groups knows, as soon as you look at the precise stem and crown boundary you get into a lot of trouble. It is the most difficult spot to pin down: taxa slip around the crown node all the time because it is hard to demonstrate they actually belong there. I was slightly concerned by your neat division between the crown and stem group, which exactly coincided with this mass extinction.

Coates: It was a cartoon; a simplification.

Budd: Could you un-cartoonify it a little?

Coates: Our argument is that the crown group is monophyletic relative to the Devonian radiation of stem group tetrapods. The crown group post-dates the Devonian–Carboniferous boundary, which is around 360 million years ago. Our estimate for the minimum age of the crown group was about 330–340 million years ago. The difference from the Devonian boundary might be a product of Romer's gap, mentioned earlier. There are a few basal stem taxa that push through (these would be conventionally thought of as lobe-finned fish, something like a lungfish, rather than as tetrapods), but these are exceptions. We also know there are stem tetrapods (of a superficially salamander-like morphology) in the Carboniferous. But what we don't find in the Carboniferous is much evidence of things like the Devonian forms, such as *Acanthostega* and *Ichthyostega*. It is worth noting here that the end of the Devonian is marked by one of the 'big five' extinction events in the fossil record, and this includes a marked turn-over in the vertebrate fauna.

Carroll: There are a couple of scapulocoracoids very like those of *Ichthyostega*.

Budd: It is the same with the Cambrian explosion, although again we have severe data limitations and can't see much of what is going on above the upper Cambrian. The major radiations are accompanied both by elevated rates of extinction as well as diversification. It's not surprising that the stem groups tend to go to the wall.

Coates: Returning to the earlier point about the significance of the East Kirton vertebrate fauna, because of the diversity of taxa present (all the main players are present), alternative tree topologies are likely to arrive at the same minimum estimate for the date of the tetrapod crown group radiation.

Budd: At the end of Jacob's paper on tinkering he has an interesting and perhaps surprising section about convergence. He says that we'd never expect the same results twice because of tinkering. As you mentioned yourself, we tend to see homoplastic effects coming again and again within the same clade. So does tinkering promote or not promote homoplasy?

Coates: That is tinkering as a process. Tinkering as a metaphor breaks down here.

Lieberman: If we go to Michael Bell's example, we know that *Pitx1* is involved in pelvic spine reduction, but there could be many ways in which this pathway might have been modified.

R Raff: The process will depend a lot on phylogenetic position. We see similar things happening in sea urchins. There are parallel events because they are all closely related genetically. If you go further out and ask why whales have the same body shape as sharks, you get a different result.

Hall: Jacob is probably thinking in terms of homology being common development, and is constraining his arguments in that way.

Bard: When you were talking about the fin fold bud, you said something about neural crest migrating it. What did you have in mind?

Coates: I had in mind skeletogenic neural crest cells migrating in to the fin bud. We have dermal fin rays out there, and as far as we know the dermal skeleton is a neural crest derivative.

Bard: It is worth noting that the parietal bone of the mouse skull is dermal, not endochondral bone and it at least is a mesodermal rather than a neural crest derivative

Hall: This is a different dermal skeleton. Your point about the median fin was interesting: the mesenchyme in there is of neural crest origin.

Coates: We know that neural crest pigment cells enter the pectoral fins of certain teleosts. There is a lovely descriptive paper by Trinkaus (1988) in which they are described going over the yolk sac and they migrate into the fin buds, where they (then) align with the fin rays.

Hall: They certainly go into the fin bud.

Duboule: There has been a lot of controversy on this. To my knowledge there is no good evidence demonstrating this, even in zebrafish where lineage tracing experiments are possible.

Hall: There is evidence that neural crest cells go into the fins and make mesen-chymal cells. There is no evidence that these cells then make fin rays.

References

Clack JA 2002a An early tetrapod from 'Romer's Gap'. Nature 418:72–76

Clack JA 2002b Gaining ground: the origin and early evolution of tetrapods. Indiana University Press, Bloomington

Holder N 1983 Developmental constraints and the evolution of vertebrate digit patterns. J Theoret Biol 104:451–471

Lebedev OA, Coates MI 1995 The postcranial skeleton of the Devonian tetrapod *Tulerpeton curtum* Lebedev. Zool J Linn Soc 114:307–348

Ruta M, Coates MI, Quicke DLJ 2003 Early tetrapod relationships revisited. Biol Rev 78:251–345

Trinkaus JP 1988 Directional cell movement during early development of the teleost *Blennius pholis*: II. Transformation of the cells of epithelial clusters into dendritic melanocytes, their dissociation from each other, and their migration into the pectoral fin buds. J Exp Biol 248:55–72

Zakany J, Fromental-Ramain C, Warot X, Duboule D 1997 Regulation of number and size of digits by posterior Hox genes: a dose-dependent mechanism with potential evolutionary implications. Proc Natl Acad Sci USA 94:13695–13700

Craniofacial variation and developmental divergence in primate and human evolution

Rebecca R. Ackermann

Department of Archaeology, University of Cape Town, Private Bag, Rondebosch 7701, South Africa

Abstract. Many questions about developmental divergence in human (and non-human primate) evolution can be fruitfully explored through investigation of the extant primate phenotype. Here I discuss two approaches that use patterns of variation in extant primates to consider hypotheses of 'tinkering' both in their own lineages, and also as applied to the fossil record of human evolution. In the first, I show how comparisons of ontogenetic morphological integration in extant humans and apes can be used to consider the developmental underpinnings of the morphological change seen in the transition from the prognathic australopith face to the relatively smaller, orthognathic *Homo* face. In the second approach, I demonstrate how studies of craniofacial variation in hybrid baboons can be used as models for considering developmental divergence in Plio-Pleistocene primates, including fossil hominins. Of particular interest is the fact that unusual non-metric dental and sutural variation in these hybrids appears to be a sensitive indicator of evolutionary developmental divergence. Future studies would profit from focusing on the breadth and especially the overlap of morphological variation among extant primate taxa in order to determine the degree to which underlying genetic similarity in functional regions, and difference in regulatory regions, explains the variable primate phenotype.

2007 Tinkering: the microevolution of development. Wiley, Chichester (Novartis Foundation Symposium 284) p 262–279

Primates are incredibly diverse. They range in body size from 25 grams to in excess of 200 kilograms, inhabit a range of environments including everything from tropical rainforests to savannas, occupy both diurnal and nocturnal niches, and display a broad range of dietary adaptations. Humans are probably the most extraordinary of all the primates, with among other things our bipedal locomotion, huge brains, complex language, and extreme dependence on tools. Yet, few of the differences in skeletal morphology seen among primates—including humans— can be considered evolutionarily 'large-scale' as defined by Carroll et al (2005), as there is relatively little: meristic variation across taxa, diversification of serially homologous parts, diversification of homologous parts, or true novelty in the functional sense (Carroll et al 2005).

Instead, the vast bulk of primate skeletal diversity is essentially variation on a theme, and as such probably derives from 'tinkering,' e.g. small-scale shifts in development (sensu Lieberman et al 2004). This is particularly true for human evolution (Carroll 2003). Our earliest ancestors diverged from the chimpanzee ancestor around 5–8 million years ago, and following this all hominins were bipedal (although the earlier hominins may have spent some portion of their time locomoting arboreally), and shared the same general skeletal organization. The emergence of our genus *Homo* was marked by a significant increase in both relative and absolute brain size, the attainment of a larger body size, and the appearance of an essentially modern body shape. Differences in skeletal morphology following this became increasingly subtle (see discussion in Lieberman et al 2004). As such, it is important to consider regulatory change as responsible for much of the morphological change seen in human evolution—particularly from the early Pleistocene onwards; unfortunately, these small-scale shifts in development can be less easy to detect than large-scale ones.

Here I will focus on how tinkering in primate evolution can be studied using a 'phenotype-to-genotype' approach (Lieberman et al 2004), where hypotheses about the developmental processes responsible for morphological divergence in the past are tested via empirical approaches that interpret patterns of extant phenotypic variation and covariation in the context of phylogenetic relationships. Such approaches are particularly useful for interpreting evolutionary change in the primate skull, and I will concentrate on this region. I will first briefly review studies of hominin developmental divergence in the Plio-Pleistocene. I will then discuss two different approaches to this problem, drawn from my own work. Both approaches are useful for elucidating the evolutionary developmental divergence of the studied extant primates themselves, but also for interpreting the fossil evidence of human evolution. In the first, I show how comparisons of patterns of variation and integration in extant humans and apes can be used to consider the developmental underpinnings of the morphological change seen in the early *Homo* face. In the second example, I consider hypotheses regarding developmental divergence in more recent *Homo* in light of studies of craniofacial variation in hybrid baboons. Finally, I will offer some thoughts on how to further test hypotheses of 'tinkering' by studying the breadth of morphological variation in extant primates.

Developmental divergence in hominin crania

Rising interest in the evolution of development in hominins has led to the recent proliferation of studies of hominin growth. However, relatively few have dealt directly with fossil material, in part because of the dearth of juvenile fossils. The bulk of these studies of hominin development have focused on the question of whether particular species show a more 'human-like' or 'ape-like' pattern and/or rate of development. They have generally (though not exclusively) been concerned with the face and dentition, both because of the relative abundance of dental remains, and because cranial remains tend to be more diagnostic and have generally formed the basis for taxonomy across hominins. A review of this literature in its entirety is beyond the scope of this manuscript (see Minugh-Purvis & McNamara 2002, Thompson et al 2003 for a recent overview), but in general these studies have focused on two groups—late Pliocene/early Pleistocene hominins, including australopiths and early *Homo*, and late Pleistocene hominins, especially Neanderthals—and have resulted in two broad conclusions. First, that earlier hominins display varied patterns of development; the best modern analogue for *Australopithecus* is probably the chimpanzee (although it does differ from the chimpanzee in some respects), while *Paranthropus* has a unique pattern of development that is nonetheless similar to modern humans in some respects (see review in Kuykendall 2003). Additionally, tooth formation times (and by extension the overall rate of development) are shorter in australopiths and early members of the genus *Homo*, relative to modern humans (Dean et al 2001). Second, while later members of the genus *Homo* share their general patterning of growth with modern humans, they still differ in having a faster rate of growth; this is especially true for Neanderthals (Ponce de León & Zollikofer 2001, Ramirez Rozzi & Bermudez de Castro 2004). Importantly, comparisons between Neanderthals and modern humans also suggest that craniofacial differences between these taxa arise ontogenetically early (Krovitz 2000, Ponce de León & Zollikofer 2001, Williams et al 2002, Zollikofer & Ponce de León 2004). Similarly, there is some evidence that the morphological differences among earlier hominins, and between them and modern analogues, appear to be set quite early in development (Dean & Wood 1984, Dean 1988, Ackermann & Krovitz 2002, McNulty et al 2006). This is particularly important for considering the genetic basis for such differences, as it suggests that we should look towards regulators of early development (e.g. early in the period of differential growth, or perhaps earlier) as the important drivers of phenotypic diversification in human evolution. These studies have not, however, provided clear insight into what these regulatory differences might be.

Morphological integration as evidence of developmental divergence

Part of the problem with relying solely on fossil material for illuminating the evolution of hominin development is the dearth of juvenile material. Neanderthals have the most complete ontogenetic series of all the hominins, and even that is rather thin for studies of development, especially in the earliest age categories. Other hominin species have few or no juvenile or infant remains. Because of this, approaches that interpret fossil evidence in the context of ape and human ontogeny play an important role in pinpointing the developmental genetic basis of phenotypic divergence in human evolution. One such approach was outlined by Lieberman and colleagues (Lieberman et al 2004). They sketched out a series of analytical steps for testing hypotheses about the basic modules of evolutionary change: (1) identify morphologically integrated regions, (2) identify differences in integration/modularization between taxa, (3) examine ontogenetic series to see if and where these changes occur, and (4) hypothesize about what underlying processes are responsible for the observed changes in morphology (Lieberman et al 2004). By using this approach to consider the phenotypic change necessary to produce the modern human cranium from an ancestral, more ape-like form, they showed that facial retraction in the human cranium is largely the product of a relatively shorter upper face, a relatively longer cranial base, and a more flexed cranial base angle, all possibly resulting from changes in the epigenetic interactions between the cranial base and the brain/face (Lieberman et al 2004).

I have used a similar tactic to consider changes in facial integration across ontogenetic series of humans, chimpanzees, bonobos and gorillas (Ackermann 2005). This work has shown that while the overall pattern and magnitude of integration is similar across these species, there are some subtle differences between them. One difference has to do with the pattern of integration. There were two main contributors to facial integration: oral and zygomatic regions. Oral integration existed early in ontogeny, and continued through adulthood in all of the species. However, the timing of the onset of zygomatic integration differed among the species. The major difference was between humans and the apes, with human zygomatic integration only occurring in late ontogeny, while in the apes zygomatic integration appears earlier. Another important result had to do with the overall magnitude of total facial integration. In adults, integration is highest in humans, followed by chimpanzees and the other two species. This is in contrast to the relatively low levels of integration in human non-adults, relative to what is seen in the apes, and suggests early developmental plasticity in the human lineage. This was corroborated by a comparison of covariance structure across ontogenetic stages, where it became clear that there is a major shift in integration late in human ontogeny compared to an earlier shift in apes, around the time of sexual maturation. While it is important to keep in mind the problems associated with sample heterogeneity in such comparisons, these results are nonetheless compelling, and may provide evidence for distinctly different developmental integration in humans.

Turning to the fossil record, let's assume for the sake of argument that the differences in integration are derived in our lineage, and that the chimpanzee condition is a better model for australopiths, and the modern human for early *Homo*. What kinds of morphological changes might be associated with these differences in integration? Figure 1 shows a plot of Procrustes scaled coordinates for the craniofacial data points used in the above study, based on sixty adult chimpanzees and sixty adult humans, as well as an adult *Australopithecus africanus* (STS 5). The shape change necessary to produce a human from a chimpanzee is primarily explained by reduction of the lower face, both in prognathism and to a lesser extent height, and an expanded, more vertically oriented middle/upper face. There is little systematic change visible in the zygomatic region, perhaps because this was specific to gorillas. The australopith (STS 5) generally aligns with the chimpanzee, particularly in the most divergent regions of the lower and middle face, confirming that the chimpanzee is likely a good model for considering this evolutionary transition.

These results support Lieberman et al's (2004) hypothesis that reduction of the face was driven by underlying developmental 'tinkering,' as evidenced by shifts in integration. But what is the underlying genetic basis for this shift in integration and decrease in facial size? Most likely, differences may be influenced (at least in part) by differences in growth hormones, possibly tied to relatively early and

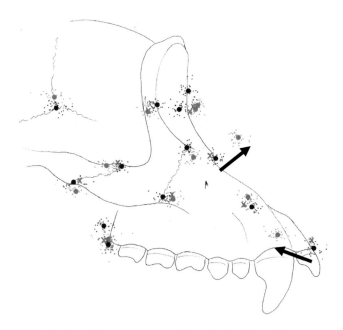

FIG. 1.   Facial landmarks of Procrustes rotated data, depicted on a chimpanzee outline. The average landmark position for each species is depicted as a large circle. Chimpanzee landmarks are shown as black circles, humans as grey circles, and *A. africanus* (STS 5) as grey Xs. The arrows indicate landmark variables which exhibit the most shape change in evolving from an ape-like, prognathic face, to the typically orthognathic face of *Homo*. Note that in the regions of greatest change the australopith aligns most closely with the chimpanzees.

sustained growth in the apes, and by correlated functional/mechanical differences—specifically those which influence the size/shape of the oral/masticatory region. This suggests that we should look towards processes which underlie facial growth, and dental development. Lieberman et al (2004 p 296) discuss how decreases in facial size might result from decreased growth during ontogeny, especially since the faces of people with growth hormone deficiency show proportionally less growth in facial dimensions (Kjellberg et al 2000). As they point out, potential mechanisms to face diminution include modifications of the growth hormone (GH)–insulin-like growth factor 1 (IGF1) axis and the thyroid hormone (TH) axis (Johnston & Bronsky 1995). The results discussed here are consistent with this.

Finally, it is important to consider the evolutionary cause of such shifts in development. One possible reason for this is that many of the functional constraints on the hominin face were released with the advent of stone tool technology, essentially allowing the early *Homo* face to vary more freely. In such a scenario, strong, ontogenetically early facial integration would be the ancestral condition across the apes, and was lost in *Homo*. This is consistent with another study that supports a hypothesis of drift as an explanation of early *Homo* facial diversity (Ackermann & Cheverud 2004).

Hybrids as detectors of developmental divergence

Another potentially fruitful approach for considering developmental divergence in primates is through analyses of the morphology of hybrids. Jernvall and Jung (2000) suggest that natural mutants with defects in cusp morphology can be informative for considering cusp development (Jernvall & Jung 2000). Here I propose that the same principle is true of natural hybrids, as presumably taxa that are extremely divergent evolutionarily would exhibit a breakdown in hybrid development. The morphological manifestation of this breakdown can help us to pinpoint the underlying units of developmental divergence. This is of obvious use for considering the genetic underpinnings of evolutionary divergence of the parental groups, but it also allows us to examine developmental divergence in the morphology of fossil primates, including human ancestors.

Current research is being conducted on a sample of yellow baboons, olive baboons, and their hybrids from the Southwest Foundation for Biomedical Research. These hybrid baboons consist of first generation and backcrossed hybrids, all of known pedigree (Ackermann et al 2006). The parental populations diverged around 160000 years ago (Newman et al 2004). One of the most interesting results coming out of this work is a high level of unusual non-metric trait variation in the hybrids (see discussion in Ackermann et al 2006). In particular, the incidence of supernumerary teeth is higher than expected in the hybrid males, and the incidence of unusual zygomaxillary sutures higher in the hybrid females.

The frequency of tooth abnormalities is particularly striking; 50% of F_1 hybrid males having at least one supernumerary fourth molar, the bulk of which are mandibular, with 90% of these bilaterally expressed. One interesting individual also exhibits full-sized maxillary supernumerary canines. Permanent, full-sized supernumerary teeth are uncommon in modern humans and other primates, with an incidence below five percent (Lavelle & Moore 1973, Rajab & Hamdan 2002), and bilateral expression of rare non-metric traits is even less common (Hallgrímsson et al 2005). Additionally, most supernumerary teeth in primates are maxillary and occur in the anterior dentition (Lavelle & Moore 1973, Rajab & Hamdan 2002). Clearly the anomalies seen here differ not just in frequency from the pure-bred populations, but also in expression (Table 1).

Developmental explanations for these anomalies likely reside in the genes responsible for the coordination of dental development, and specifically dental arch segmentation. The distomolars seen here might result from finer subdivisions of the molar dental field or rather a longer molar dental field divided into normal sized segments. To test this, I measured four variables on each individual male hybrid: the maximum length of the first to third molar at the dento-enamel junction in each of the four dental quadrants. Quantitative differences among male individuals with supernumerary teeth and those without suggest that the alveoli (sans distomolars) are normal sized (MANOVA; $n = 17$; $P = 0.081$), although univariate tests indicate that the length of the molar tooth row in the left mandible is significantly smaller in the hybrids with distomolars ($P = 0.038$). Additionally visual examination shows that the morphology of the first three molars in indi-

TABLE 1 Supernumerary teeth in baboons

Taxon	ID	Sex	Supernumerary tooth	Bilateral?	Full-sized?
Olive	W136	F	Maxillary	Y	Y
Olive	W206	F	Maxillary	N	N
Olive	W235	M	Maxillary	N	N
F1 hybrid	W124	F	Mandibular	Y	Y
F1 hybrid	W114	M	Mandibular	Y	N(left)/Y(rt)
F1 hybrid	W135	M	Mandibular	Y	N
F1 hybrid	W155	M	Mandibular	Y	Y
F1 hybrid	W159	M	Maxillary & Mandibular	Y	Y
F1 hybrid	W18	M	Mandibular	Y	Y
F1 hybrid	W199	M	Mandibular	N	N
F1 hybrid	W68	M	Mandibular	Y	N
F1 hybrid	W97	M	Maxillary & Mandibular	Y	Y
F1 hybrid	W	M	Canine	Y	Y

Data from Ackermann et al (2006).

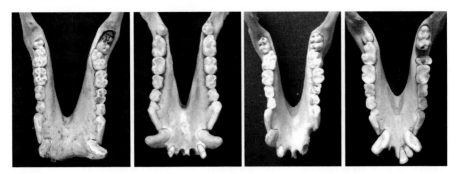

FIG. 2. Examples of supernumerary mandibular teeth in hybrids, from left to right: W18, W68, W159, W97. Note that when the teeth are full sized, they most closely resemble the morphology of a typical third molar, although there is a lot of variation in overall morphology.

viduals with distomolars is normal, while the morphology of distomolars, when full sized, most often resembles the third molar, though this is variable (Fig. 2). This pattern suggests that the supernumerary teeth result from a longer molar dental field, rather than finer subdivisions of a normal sized dental field, possibly due to an extension of tooth morphogenesis in these hybrids.

The mechanism of the development of supernumerary teeth in humans is still unknown, although recent research is consistent with this hypothesis, suggesting that an increased number of molars may be tied to the lack of a termination process. For example, it is possible that supernumerary teeth might develop from remnants of the dental lamina which have not dissolved (Wang 2004). Keeping in mind that the likelihood of a single-gene effect is small, one candidate for such a mechanism is *Runx2*; recent studies indicate that *Runx2* is both required for advancing tooth morphogenesis and histodifferentiation (D'Souza et al 1999), and also plays an inhibitory role in the formation of successive dentition (Aberg et al 2004). Interestingly, in the absence of *Runx2* activity in mouse mutants, the lower molars seem to be more severely affected than the upper molars (Wang et al 2005). Another possible candidate is Ectodysplasin; ectodysplasin–Edar signalling performs several roles in ectodermal organ development, controlling the initiation of organs, as well as their morphogenesis and differentiation (Mustonen et al 2003). Importantly, supernumerary teeth and other dental abnormalities occur in mice with K14-EDA-overexpression (Mustonen et al 2003).

It is interesting to consider how this relates back to the ultimate cause of baboon divergence, i.e. evolution. One possibility is that underlying developmental mechanisms have drifted between the two species and that hybridization therefore produces a perturbed morphology. It is also possible that the genetic divergence seen here is the result of selection, particularly since abnormal traits that appear in hybrid offspring but not in parental animals are considered the result of mixing two separately co-adapted genomes (Falconer & Mackay 1997, Vrana et al 2000). But if this divergence indeed results from selection, what precisely is the (divergent) adaptation? There is no clear reason for postulating a selective advantage for differences in dental development among baboons. However, it is quite possible that what we are seeing is essentially a 'spandrel,' as this difference might result from selection on other aspects of morphology that are targeted by these regulatory genes. For example, overexpression of K14-EDA also affects the growth of other ectodermal organs, resulting in supernumerary mammary glands, abnormal hair structure, unusually long hair and nails, stimulated sweat gland function and enlarged sebaceous glands (Mustonen et al 2003). In animals such as the baboon that are geographically widespread across latitudinally distinct ecosystems, there are clear phenotypic differences in hair thickness, coat colour, and other ectodermally derived tissues. Quite possibly, selection on these aspects of morphology has produced small differences in the timing of development in this system.

Finally, how does this inform our interpretations of the fossil record? First of all, it must be established that this type of trait variation is not unique to our sample. Although the effects of hybridization on dental morphology in mammals are not well studied, there are other documented cases of supernumerary teeth and dental abnormalities in hybrids. Supernumerary teeth, including distomolars and one incidence of double canines, have been reported in individuals collected near Lake Malawi, in a region of known sympatry between chacma and olive baboons (Freedman 1963, Hayes et al 1990). Supernumerary teeth are also common in both recent and late Pleistocene (>120000 years) ground squirrel hybrids (Goodwin 1998), while dental abnormalities in a monodontid skull have also been attributed to hybridization—in this case between the closely related belugas and narwhals (Heide-Jørgensen & Reeves 1993). These studies suggest that the unusual dental signatures of developmental divergence seen in the baboons are probably characteristic of inter-specific mammalian hybridization more broadly. This means that we should be looking for similar signs of developmental breakdown in the hominin fossil record, in the form of a prevalence of dental and osteological anomalies, and particularly rare bilateral anomalies, as one way to detect developmental divergence (Ackermann et al 2006). The only such candidate currently in the hominin fossil record is the Flores hominin; the type skull of *Homo floresiensis* exhibits rotated maxillary premolars, which is an unusual dental anomaly not recorded in any other hominin (Brown et al 2004). Whether this indicates hybridization between two distinct genetic populations, or is instead a signature of the breakdown of development resulting from other phenomena—such as island dwarfism, pathological development, or simply extremely reduced facial size—is not clear.

Future possibilities: examining ranges of variation in populations

While it is true that much will be learned from comparing the genomes of humans and chimpanzees (and other primates as they are sequenced), the connection between the genotype and phenotype remains elusive for the vast bulk of morphological traits. As has been shown here, a focus on phenotypic traits of interest may provide guidelines for targeted study of genomic variation. More importantly, it allows for a direct connection to the processes of evolution that might be guiding phenotypic change, and potentially may provide us with insight into ways in which selection for some phenotypes may result in the correlated evolution of others. One of the things that has stood out in considering these studies is that the traits under consideration are not novel; even double canines have been reported once before in the literature. This suggests that changes in regulatory regions are unlocking 'cryptic' variation, and perhaps closer scrutiny of ranges of intraspecific variation in primates could provide further insight into evolutionary 'tinkering.' When ranges of variation overlap—especially when a trait that is atypical in one population is common in another—this indicates underlying genetic similarity in functional regions and differences in regulatory regions (Carroll 2005). There is overwhelming evidence for just this sort of distribution of morphology throughout primate and particularly human evolution; it begs further exploration.

Acknowledgements

The inspiration for much of this work has emerged from conversation and collaboration with my teacher, mentor, and now colleague, Jim Cheverud. I am very grateful for his support and his friendship. Special thanks are also due to Dan Lieberman and Brian Hall for organizing this symposium, to Jim Cheverud and Benedikt Hallgrímsson for acting as referees for this manuscript, and to the Novartis Foundation for financial support. Research was funded by grants from the National Research Foundation of South Africa and the University Research Committee of the University of Cape Town.

References

Aberg T, Cavender A, Gaikwad J et al 2004 Phenotypic changes in dentition of Runx2 homozygote-null mutant mice. J Histochem Cytochem 52:131–139
Ackermann RR 2005 Ontogenetic integration of the hominoid face. J Hum Evol 48:175–197
Ackermann RR, Krovitz GE 2002 Common patterns of facial ontogeny in the hominid lineage. Anat Rec 269:142–147
Ackermann RR, Cheverud JM 2004 Detecting genetic drift versus selection in human evolution. Proc Natl Acad Sci USA 101:7946–17951
Ackermann RR, Rogers J, Cheverud JM 2006 Identifying the morphological signatures of hybridization in primate and human evolution. J Hum Evol 51:632–645
Brown P, Sutikna T, Morwood MJ et al 2004 A new small-bodied hominin from the late Pleistocene of Flores, Indonesia. Nature 431:1055–1061
Carroll SB 2003 Genetics and the making of *Homo sapiens*. Nature 422:849–857

Carroll SB, Grenier JK, Weatherbee SD 2005 From DNA to diversity: molecular genetics and the evolution of animal design 2nd edn. Blackwell Publishing, Oxford

Dean C, Leakey MG, Reid D et al 2001 Growth processes in teeth distinguish modern humans from *Homo erectus* and earlier hominins. Nature 414:628–631

Dean MC 1988 Growth processes in the cranial base of hominoids and their bearing in morphological similarities that exist in the cranial base of *Homo* and *Paranthropus*. In: Grine FE (ed), Evolutionary history of the 'robust' Australopithecines. Aldine de Gruyter, New York p 107–112

Dean MC, Wood BA 1984 Phylogeny, neoteny and growth of the cranial base in hominids. Folia Primatol 43:157–180

D'souza RN, Aberg T, Gaikwad J et al 1999 Cbfa1 is required for epithelial-mesenchymal interactions regulating tooth development in mice. Development 126:2911–2920

Falconer DS, Mackay T 1997 Introduction to quantitative genetics. Dover Publications, New York

Freedman L 1963 A biometric study of *Papio cynocephalus* skulls from Northern Rhodesia and Nyasaland. J Mammal 44:24–43

Goodwin HT 1998 Supernumerary teeth in Pleistocene, recent, and hybrid individuals of the *Spermophilus richardsonii* complex (Sciuridae). J Mammal 79:1161–1169

Hallgrímsson B, Donnabháin BÓ, Blom DE, Lozada MC, Willmore KT 2005 Why are rare traits unilaterally expressed?: trait frequency and unilateral expression for cranial nonmetric traits in humans. Am J Phys Anthrop 128:14–25

Hayes VJ, Freedman L, Oxnard CE 1990 The taxonomy of savannah baboons: an odontomorphometric analysis. Am J Primatol 22:171–190

Heide-Jørgensen MP, Reeves RR 1993 Description of an anomalous monodontid skull from West Greenland: A possible hybrid ? Mar Mamm Sci 9:258–268

Jernvall J, Jung H-S 2000 Genotype, phenotype, and developmental biology of molar tooth characters. Yearb Phys Anthropol 43:171–190

Johnston MC, Bronsky PT 1995 Prenatal craniofacial development: new insights on normal and abnormal mechanisms. Crit Rev Oral Biol Med 6:368–422

Kjellberg H, Beiring M, Albertsson Wikland K 2000 Craniofacial morphology, dental occlusion, tooth eruption, and dental maturity in boys of short stature with or without growth hormone deficiency. Eur J Oral Sci 108:359–367

Krovitz GE 2000 Three-dimensional comparisons of craniofacial morphology and growth patterns in Neandertals and modern humans. Ph.D. Dissertation, Johns Hopkins University

Kuykendall KL 2003 Reconstructing australopithecine growth and development: What do we think we know? In: Thompson JL, Krovitz GE, Nelson AJ (eds) Patterns of growth and development in the Genus *Homo*. Cambridge University Press, Cambridge p 191–218

Lavelle CL, Moore WJ 1973 The incidence of agenesis and polygenesis in the primate dentition. Am J Phys Anthropol 38:671–679

Lieberman DE, Krovitz GE, McBratney-Owen B 2004 Testing hypotheses about tinkering in ther fossil record: the case of the human skull. J Exp Zoolog B Mol Dev Evol 302B:284–301

McNulty KP, Frost SR, Strait DS 2006 Examining affinities of the Taung child by developmental simulation. J Hum Evol 51:274–296

Minugh-Purvis N, McNamara KJ 2002 Human evolution through developmental change. The Johns Hopkins University Press, Baltimore and London

Mustonen T, Pispa J, Mikkola ML et al 2003 Stimulation of ectodermal organ development by Ectodysplasin-A1. Dev Biol 259:123–136

Newman TK, Jolly CJ, Rogers J 2004 Mitochondrial phylogeny and systematics of baboons (*Papio*). Am J Phys Anthropol 124:17–27

Ponce De León MS, Zollikofer CPE 2001 Neanderthal cranial ontogeny and its implications for late hominid diversity. Nature 412:534–538

Rajab LD, Hamdan MAM 2002 Supernumerary teeth: review of the literature and a survey of 152 cases. Int J Paediatr Dent 12:244–254

Ramirez Rozzi FV, Bermudez De Castro JM 2004 Surprisingly rapid growth in Neanderthals. Nature 428:936–939

Thompson JL, Krovitz GE, Nelson AJ 2003 Patterns of growth and development in the Genus *Homo*. Cambridge University Press, Cambridge

Vrana PB, Fossella JA, Matteson P, Del Rio T, O'Neill MJ, Tilghman SM 2000 Genetic and epigenetic incompatabilities underlie hybrid dysgenesis in *Peromyscus*. Nat Genet 25: 120–124

Wang X, Aberg T, James M, Levanon D, Groner Y, Thesleff I 2005 Runx2 (Cbfa1) inhibits Shh signaling in the lower but not upper molars of mouse embryos and prevents the budding of putative successional teeth. J Dent Res 84:138–143

Williams FL, Godfrey LR, Sutherland MR 2002 Heterochrony and the evolution of Neandertal and modern human craniofacial form. In: Minugh-Purvis N, Mcnamara KJ (eds), Human evolution through developmental change. The Johns Hopkins University Press, Baltimore and London

Zollikofer CPE, Ponce De León MS 2004 Kinematics of cranial ontogeny: heterotopy, heterochrony, and geometric morphometric analysis of growth models. J Exp Zoolog B Mol Dev Evol 302B:322–340

DISCUSSION

Wagner: Can you outline how you get from measurements of integration to ontogenetic causes? How does this work exactly?

Ackermann: By identifying regions that are integrated differently through an ontogenetic series, one can pinpoint modules that have changed, assumedly under the influence of regulatory genes. Perhaps more interestingly, if you're moving through time and you see change in the covariance structure, this might be indicative of the action of selection, and if there is such a shift, then presumably selection would be targeting the regulatory genes that would act on a new unit. This might allow you to identify regulatory genes that are driving evolutionary changes in morphology.

Wagner: Benedikt Hallgrimsson showed us earlier that the degree of integration depends on how much variation there is in different processes. It could be that there is no developmental change other than some processes being more variable in the population being studied than others.

Lieberman: My model was a little bit different than that. I wasn't suggesting that it is measures of integration itself that help identify what changed, but rather that in order to generate hypotheses about where selection has acted, we first need to figure out where the units are that have changed in the first place. We can first identify modules—units that have some kind of coherence of structure—then we can figure out how the interactions between these units have then changed. This gives a range of hypotheses we can begin to start narrowing down. There is a problem with variation: increased variation will increase covariation. I would agree that the measure of integration itself may not give any information about what has changed.

Wagner: Is the goal to find out what selection has worked on? Or is the goal to find the developmental knob that selection has turned?

Lieberman: We need to do both. If we don't know where development has acted, we don't know where selection has acted. We don't know what is coming along for the ride, and what was actually selected for in the first place. One of the problems with skulls is that they are so integrated: if you change one region you change another region. You may see the change in one region and you could falsely infer selection. It is not easy.

R Raff: I was curious about your use of hybrids. It seems reasonable to look at where something disintegrates a bit in a hybrid in order to look for possible candidates, but I am puzzled about your statement about looking at the fossil record for these kinds of anomalies. The sampling is tiny. Each species perhaps has a species duration of a million years, and during that time each is presumably as well integrated as we are. If there are evolutionary changes taking place where there is a transition, you are unlikely to find these transitional individuals. Why would you expect to find these anomalies in the fossil record?

Ackermann: There has been an enormous amount of discussion in the literature about potential for hybridisation and contact across the Plio-Pleistocene, both among recent populations such as Neanderthals and modern humans, and also further back in time between early chimpanzee and human ancestors. Recently, especially in the discussion over the relationship between modern humans and Neanderthals, it has become commonplace for researchers to refer to species as developmentally very different from one another, and yet still interbreeding at low frequencies. What I am getting at here is that if such species were developmentally *that* different, either they were unlikely to be hybridising, or we should see some sort of sign of this hybridization. As far as finding them in the fossil record, that is a very different question. Our fossil record is good by some standards, but still quite small as far as species sample sizes are concerned. This is why we are saying that our ability to test for population-level phenomena such as increased trait frequencies is virtually non-existent. However, if you do find something that is clearly indicative of developmental breakdown in the phenotype, such as rare bilateral dental anomalies, it is a bit of a red flag.

Hallgrimsson: Your finding that the zygomatic covariance in humans is reduced is interesting. Do you think that this is because humans have a smaller masseter and temporalis so they are generating less variance?

Ackermann: That was my thinking, especially relative to chimpanzees

Hallgrimsson: The hybrids or parental species differ in facial length. Is there a relationship between variation in the length of the mandible and the occurrence of these supernumerary teeth?

Ackermann: Some of these hybrid males are huge, so I also thought there would be a correlation between extra teeth and a longer alveolus, but there doesn't seem to be. There is a lot of size variation across these hybrids, and some really big hybrids have supernumerary teeth, and some don't. I also looked at heterosis in the hybrids, and there was surprisingly little (Ackermann et al 2006).

Thesleff: I wanted to ask about the supernumerary teeth. In humans they are very rare. The only syndrome is cleidocranial dysplasia, which is caused by heterozygous loss of function of the *Runx2* gene. These people have extra posterior molars, but even more typical is that they have supernumerary replacement teeth, so they may have an almost complete third dentition. Have you looked at this in baboons?

Ackermann: We haven't, but they are now being CT scanned, so this is something we could look at.

Thesleff: In humans, missing teeth are common. Around 8% of people have tooth agenesis. How common is this in primates?

Ackermann: In these baboons I didn't see any missing teeth aside from attrition.

Cheverud: It hasn't been documented much in primates other than humans.

Weiss: Humans may be a transitional species in terms of diet, so we could be on the way to losing teeth.

Cheverud: Our teeth are getting smaller.

Lieberman: Yes, even over the last 100 years. There are clear environmental effects.

Oxnard: Would there be any interest in looking at the Sulawesi macaques, where there is a star-shaped island with several groups. A lot of the hybrid zones are wide, while others are narrow.

Ackermann: They are one of the best-studied hybrid groups among primates, but their skeletal morphology has not been well-studied. Most of the work done on them has been concerned with heterosis and overall body size.

Stern: Are you using the skeletal malformation simply to infer hybridization, or are you trying to say something more about developmental integration of these skulls from the malformations?

Ackermann: We know that these baboons are hybrids. As far as applying these results to considering the fossil record, one reason to do this is to try and identify whether hybridization has occurred. Clear breakdown in the co-ordination of early development is quite possibly an indication of hybridization.

Stern: For many years it has been known that when closely related *Drosophila* species (a couple of million years) are crossed, some of the most highly evolution-arily conserved bristles on the dorsal surface are lost (Sturtevant 1920, Skaer 2000). You would have thought that the mechanisms that pattern these bristles would have been conserved because flies have been producing bristles in these positions for hundreds of millions of years. This suggests that whatever the developmental mechanism is that generates the bristle pattern, it is changing rather rapidly, even though the phenotype is quite conserved. This makes it difficult to make inferences from abnormalities seen in hybrids to how the developmental system is evolving.

Ackermann: So you don't think that the fact that these abnormalities exist, presumably because of changes in a particular gene network, is indicative of divergence in the two parental populations?

Stern: I'm saying that there is an alternative hypothesis. Clearly the underlying mechanisms must have evolved because when the genes from different species are put together in a hybrid something goes wrong. But this doesn't mean that those are the mechanisms producing the divergence you are interested in.

Cheverud: There is very little divergence except in size among those species.

Stern: It is unclear to me how you use this information to define anything about development. When the bristles are being lost, it is likely that exactly the same molecules generate the pattern of bristles in the different species. Perhaps there are quantitative shifts in the way they are talking to one another when they define the bristle pattern.

Ackermann: That is still telling you about divergence in that system, isn't it?

Stern: It is telling you about neutral changes in the relative interactions between genes that end up producing exactly the same phenotype in their original species.

Wilkins: There is an alternative way to view this. The components are co-evolving in the different species, in part because they are also components of other developmental processes. But this means that for the specific point of bristle development, these components within the network are fairly well matched, but when you make a hybrid cross you mismatch the components. What is seen in terms of the loss of bristles is in a sense artefactual; it does indicate the existence of evolution of components, but it is not necessarily deeply significant about evolution of underlying mechanisms. I suspect the supernumerary teeth phenomenon could be artefactual, as if you have introduced some small mutations into the network which destabilize it and lead to supernumerary teeth.

Ackermann: It is only useful in the context of thinking about what we know about the morphological divergence of baboons. We know that they vary primarily in body size and coat colour, and it is reasonable to assume that these differences were selected for, though of course we can't rule out random processes of divergence.

Hall: What I thought you were going to use the data for is to say that this is a sort of atavistic character; a reflection of something that ancestors had, and that you would then look at which fossil ancestors had this tooth.

Ackermann: We don't have that tooth in fossil baboon ancestors.

Lieberman: In this meeting we have had a number of examples about hybrids. If you get a build up of mutations between species, are hybrids from more distantly related taxa qualitatively or quantitatively different in terms of developmental disruption?

Bell: Another interesting set of studies concerns salmon in the Pacific Northwest (Gharrett & Smoker 1991, Gharrett 1999). They have very strict alternate year breeding. There are separate populations that are occupying the stream in alternate years, presumably experiencing the same conditions over the long run. If hybrids are made from the odd year and even year populations there are serious problems. These are populations that have only been in these streams for a few thousand years. They look alike but the hybrids break down.

Weiss: Should purely additive effects generate covariance?

Cheverud: I am thinking more of a morphological integration, not necessarily the direct contact of genes. In this case epigenetic interactions between tissues produce the covariation.

Weiss: Supposing you have 9 chromosome regions that give you a signal in a map. Do you imagine that somewhere in each of these regions are genes that have to have direct ligand binding?

Cheverud: I would not as there are many indirect mechanisms connecting parts of developmental processes. In some instances, I would expect ligand–receptor binding to be a potential cause of epistatic interactions among those genes. Such interactions are known to be caused by ligand-receptor binding variability between forms of ApoE and the LDL receptor, for example. I would presume that with an epistatic interaction there must be an interaction involving the pathways that they are part of, whether it's a direct physical interaction or a network-based interaction. I would expect one or the other. But I wouldn't expect this simply from the observation of morphological integration.

Weiss: So from the morphological integration you don't draw conclusions about what it implies.

Cheverud: It implies commonality of development effect.

Lieberman: Has that hypothesis ever been tested?

Cheverud: No. The part I skipped through in my presentation was concerned with how epistasis modifies the pleiotropy. So epistasis does affect morphological integration but it is a relatively poorly understood process.

Weiss: Purely additive effects wouldn't generate covariance.

Cheverud: They do. But so does epistasis. Epistasis causes covariation, as does additivity, they act together. The causes of covariation are not separable in that way.

Hallgrimsson: If you have strong selection it can reduce both variance and covariance. You can reduce integration under strong selection.

Bell: You referred on a few occasions to distinguishing selection and drift. What were your criteria?

Ackermann: The results of a previous study indicate the possibility of distinctly different ontogenetic integration in humans versus other great apes, and specifically shifts in covariance structure at different points in ontogeny (Ackermann 2005). Assumedly relative changes in the covariance structure projected over evolutionary time indicate aspects of the morphology that are being selected for. When I was referring specifically to drift and selection, it was in the context of other work where my colleagues and I have used theoretical expectations that between and within group variation in diverging lineages should be proportional to each other under a model of drift (Ackermann & Cheverud 2002, 2004).

Wagner: My impression is that a lot of the inferences you are drawing are not well understood. We know very little about how covariance matrices evolve, and what influences them, and how they change in even simple population genetic scenarios. The main thing that makes me uneasy is that the inference links that you use are entirely hypothetical. You don't even have a theoretical model supporting the inferences.

Ackermann: Fair enough. But the basic pattern of integration is the same across all the primates that have been studied, and yet there is a clear difference in ontogenetic covariance patterning between the apes and humans. Assumedly this indicates some evolutionary shift in development, though this clearly needs to be studied much more.

Cheverud: The ways specific kinds of selection would modify the covariance structure is not specified. It might be very dependent on the form of the selection. There may not be a general solution because it may depend on the details of the selection.

Wagner: And the genetic architecture.

Cheverud: One could test it against hypotheses of drift where there is a better idea of what the behaviour is supposed to be. If it is not consistent with drift you may be able to infer that selection was occurring.

Lieberman: Benedikt Hallgrimsson's data yesterday from the mutant mice showed that integration changed dramatically when he messed around with the genetic architecture. The mechanism that underlies this is still utterly unknown.

Cheverud: The theory to make the connection is not there; that's true. It's more of a statement that the differences are not due to genetic drift rather than a specific model of selection.

R Raff: I wonder if there are helpful model systems.

Lieberman: That's an interesting point. This has been a general trend: in domesticated animals there is a reduction in snout length and often an increase in brain size. However, it turns out that this pattern breaks down if you look closely enough. I have been analysing Belyaev's silver foxes. This is one of the most controlled domestication experiments ever done. I was expecting to analyse these skulls and see snout reduction. Interestingly enough, they have all the characteristics we see in domesticated species—floppy ears, curled tails and so on—but none of those craniofacial changes. This highlights the problem of analogy in evolution.

R Raff: In the case of dogs that have reduced muzzles it may be possible to look at particular genes that are involved.

Lieberman: There are people looking at Runx2 in primates.

Cheverud: There are people working on dogs, especially with the cranial form of hybrids, with long face and short face dogs, and the incredible variety seen in hybrids. The variation is high relative to the parental variation. It is also not integrated well.

Bell: Have you seen the dog farm hybrids from Cornell? The F2 phenotypes are all over the place; they are very badly integrated.

Cheverud: That is what I was referring to.

References

Ackermann RR 2005 Ontogenetic integration of the hominoid face. J Hum Evol 48:175–197

Ackermann RR, Cheverud JM 2002 Discerning evolutionary processes in patterns of tamarin (genus Saguinus) craniofacial variation. Am J Phys Anthropol 117:260–271

Ackermann RR, Cheverud JM 2004 Detecting genetic drift versus selection in human evolution. Proc Natl Acad Sci USA 101:17946–17951

Ackermann RR, Rogers J, Cheverud JM 2006 Identifying the morphological signatures of hybridization in primate and human evolution. J Hum Evol 51:632–645

Gharrett AJ, Smoker WW 1991 Two generations of hybrids between even- and odd-year pink salmon (Oncorhynchus gorbuscha): a test of outbreeding depression? Can J Fish Aquat Sci 48:1744–1749

Gharrett AJ, Smoker WW, Reisenbichler RR, Taylor SG 1999 Outbreeding depression in hybrids between odd- and even-broodyear pink salmon. Aquaculture 173:117–129

Skaer N, Simpson P 2000 Genetic analysis of bristle loss in hybrids between *Drosophila melanogaster* and *D. simulans* provides evidence for divergence of cis-regulatory sequences in the achaete–scute gene complex. Dev Biol 221:148–167

Sturtevant AH 1920 Genetic studies on *Drosophila simdans*. I. Introduction. Hybrids with *D. melanogaster.* Genetics 5:488–500

Contributor Index

280

Subject Index